Chemicotrope Impeder for Non-virulent Ferrostal Oxivaporation and Coativum Clingium

M. Tova

Chemicotrope Impeder for Non-virulent Ferrostal Oxivaporation and Coativum Clingium

First Edition October 2025

Written by M. Tova

Table of Contents

TABLE OF CONTENT

CHAPTER NO.	TITLE	PAGE NO.
I	**INTRODUCTION**	
	1.1 Introduction to Corrosion	1
	1.1.1 Historical background	1
	1.1.2 Consequences of corrosion	2
	1.1.3 Losses due to corrosion	2
	1.1.4 Definition of corrosion	4
	1.1.5 Chemistry of corrosion	5
	1.1.6 Factors influencing Corrosion	6
	1.1.7 Classification of corrosion	8
	1.1.8 Corrosion prevention	11
	1.2 Introduction to Corrosion inhibitors	12
	1.2.1 Types of inhibitors	12
	1.2.2 Mechanism of inhibition	15
	1.3 Introduction to corrosion Monitoring techniques	15
	1.3.1 Need for corrosion Monitoring techniques	15
	1.3.2 Corrosion Monitoring techniques	16
	1.4 Paints	21
	1.5 Introduction to computational chemistry	26
	1.5.1 Advantages of computational studies	27
	1.5.2 Overview of computational chemistry	27
	1.5.3 Theory of quantum chemistry	28
	1.5.4 Model chemistry	29

CHAPTER NO.	TITLE	PAGE NO.
II	**REVIEW OF LITERATURE**	
	2.1 Mild Steel	34
	2.2 Aluminium	42
	2.3 Quantum chemical studies	44
	2.4 Drugs	60
	2.5 Paint	68
	2.6 Coating	70
III	**MATERIALS AND METHODS**	
	3.1 Selection of Materials	88
	3.1.1 Selection of inhibitor	88
	3.1.2 Selection of metal	91
	3.1.3 Preparation of acid solution	92
	3.1.4 Preparation of inhibitor	92
	3.2 Methods Adopted	92
	1.2.1 Weight loss measurements	92
	3.2.2 Electrochemical measurements	92
	3.2.3 Preparation of epoxy coating	93
	3.2.4 Electrochemical measurements of the epoxy coating	94
	3.3 Theoretical methods	94
IV	**INVESTIGATION ON THE CORROSION INHIBITOR SYSTEM WITH WEIGHT LOSS MEASUREMENTS**	
	4.1 Weight loss studies	101
	4.1.1 Effect of concentration on the I.E(%) of the inhibitors	101

CHAPTER NO.	TITLE	PAGE NO.
	4.1.2 Effect of immersion time on the I.E(%) of the compounds	105
	4.1.3 Effect of temperature on the I.E(%) of the compounds	106
	4.1.4 Comparision of the corrosion inhibition behavior of the studied inhibitor	111
	4.1.5 Adsorption isotherm	112
V	**EVALUATION OF CORROSION POTENTIAL OF THE STUDIED DRUGS BY ELECTROCHEMICAL METHODS**	
	5.1 Electrochemical impedance Spectroscopic measurements	125
	5.1.1 BEX	125
	5.1.2 CIL	127
	5.1.3 MON	128
	5.1.4 PYR	130
	5.2. Potentiodynamic Polarization Studies	131
	5.2.1 BEX	132
	5.2.2 CIL	133
	5.2.3 MON	135
	5.2.4 PYR	136
VI	**THEORETICAL INVESTIGATION ON THE INHIBITIVE PROPERTY OF THE STUDIED DRUGS**	
	6.1 Structure of the inhibitor	141
	6.2 Optimized structure of the non-protonated species	142
	6.3 Molecular properties of the non-protonated species.	145

CHAPTER NO.	TITLE	PAGE NO.
	Global reactivity parameters	145
	i) Frontier molecular orbital energy	145
	ii) Energy gap	147
	iii) Dipolemoment	147
	iv) Ionization potential and Electron affinity	148
	v) Softness and Hardness	148
	vi) Electronegativity	149
VII	**ELECTROCHEMICAL INVESTIGATION ON THE PROTECTIVE DRUG/EPOXY COMPOSITES IN 3.5 % NaCl**	
	7.1.Paint composite studies	165
	7.1.1. Immersion test	165
	7.1.2 Electrochemical studies	167
VIII	**SUMMARY AND CONCLUSION**	
	8.1. Weight loss measurements	179
	8.2. Electrochemical studies	179
	8.3. Quantum chemical studies	180
	8.4. Paint composite electrochemical studies	180

Chapter I

CHAPTER I

INTRODUCTION

1.1 Introduction to Corrosion

1.1.1 Historical Background

Corrosion, commonly known as rust, deteriorates materials and shortens their lifespan. The Roman philosopher Pliny (AD 23–79) first described iron degradation in Ferrum Corrumpitar, and Robert Boyle later studied its causes in Mechanical Origin of Corrosiveness. Michael Faraday established the quantitative link between chemical reactions and electric currents, forming the basis for calculating corrosion rates. Whitney (1903) provided a framework for electrochemical control, while Schönbein (1836) showed that iron could be passivated by air exposure. Modern understanding of corrosion has been advanced by researchers like Evans, Uhlig, and Fontana. Corrosion laboratories worldwide continue to develop this field as a multidisciplinary science encompassing chemistry, materials science, and engineering Due to these need, corrosion engineers around the world organize into corrosion societies to share their experience and discoveries to name a few like,

- ***NACE International*** claims to be the worldwide society with universally accepted standards.
- ***European Federation of Corrosion (EFC)*** an association of individual European country corrosion societies.
- ***The Indiana Corrosion Society*** a technical committee in the state of Indiana.
- ***The Corrosion Science Society of Korea*** is a technical society devoted exclusively to corrosion and corrosion prevention.

The Mission of these organizations is ***"To protect people, assets and the environment from corrosion"*** and to serve as a as the premier authority for corrosion control solutions.

1.1.2 Consequences of Corrosion

Some important consequences of corrosion are summarized below:

- ***Plant shutdown***: Shutdown of nuclear plants, process plants, power plants and refineries may cause severe problems to industry and consumers.
- ***Loss of products***: Leaking containers, storage tanks, water and oil transportation lines and fuel tanks cause significant loss of product and may generate severe accidents and hazards.
- ***Loss of efficiency***: Insulation of heat exchanger tubings and pipe lines by corrosion products reduces heat transfer and piping capacity.
- ***Contamination***: Corrosion products may contaminate chemicals, pharmaceuticals, dyes, packaged goods, etc.
- ***Nuclear hazards***: The Chernobyl disaster is a continuing example of transport of radioactive corrosion products in water, fatal to human, animal and biological life.

1.1.3 Losses due to Corrosion

In general, corrosion costs in most countries are in the range of 3 – 5 % of the national gross domestic product. The losses due to corrosion may be direct loss or indirect loss as listed below:

Losses due to corrosion	Type	Examples
	Direct Loss	Over - design
		Cost of repair/ replacement
		Cost of anti-corrosive painting
		Other protection methods
	Indirect Loss	Contamination of product
		Loss of valuable products
		Loss of production
		Damage of equipment

In a study of corrosion cost conducted recently in USA, the direct corrosion cost was estimated to be around 1.4 trillion US dollars, approximately 3 – 5 % of the national gross domestic product (GDP). This figure reflect only the direct cost of corrosion and does not include the environmental damage, waste of resources, loss of production, or personal injury resulting from corrosion.

In UK, the corrosion cost is estimated to be 4–5% of the GNP. In Japan, the cost of corrosion is estimated to be 5258 trillion Yen per year. For most industrialized nations, the average corrosion cost is 3.5–4.5% of the GNP [4].

Corrosion cost in India

In India, it is know that 3% of the GDP eroded in corrosion every year, which is equal to the revenue contributed by the Indian Railways. The loss will be up to 6.5% if indirect cost of corrosion is taken into account. This cost was determined by analyzing 26 industrial sectors in which corrosion was known to exist and these sectors were divided among five major categories (**Figure 1.1**).

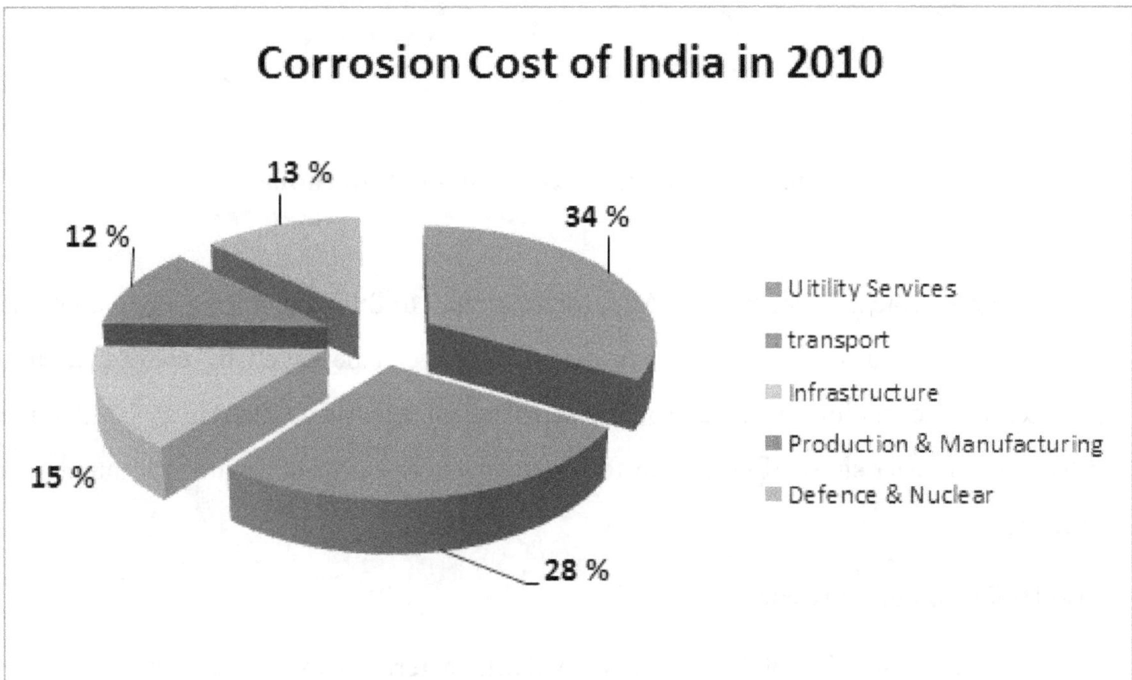

Figure 1.1 Sector wide corrosion cost of India, 2010

There exists few published data on corrosion cost of India, due to limited research on this topic. The earliest work was done by Rajagopalan in 1960 by CSIR. After that there were no such reports for Indian corrosion cost. **Figure 1.2** report the collected corrosion cost (in USD) for the period of 1960-2014 from different sources (Khanna 2002, Financial express 2004, CORCON 2007, etc.). Figure 3 clearly proves the increases in the loss due to corrosion which has a serious impact on the country economy [5]

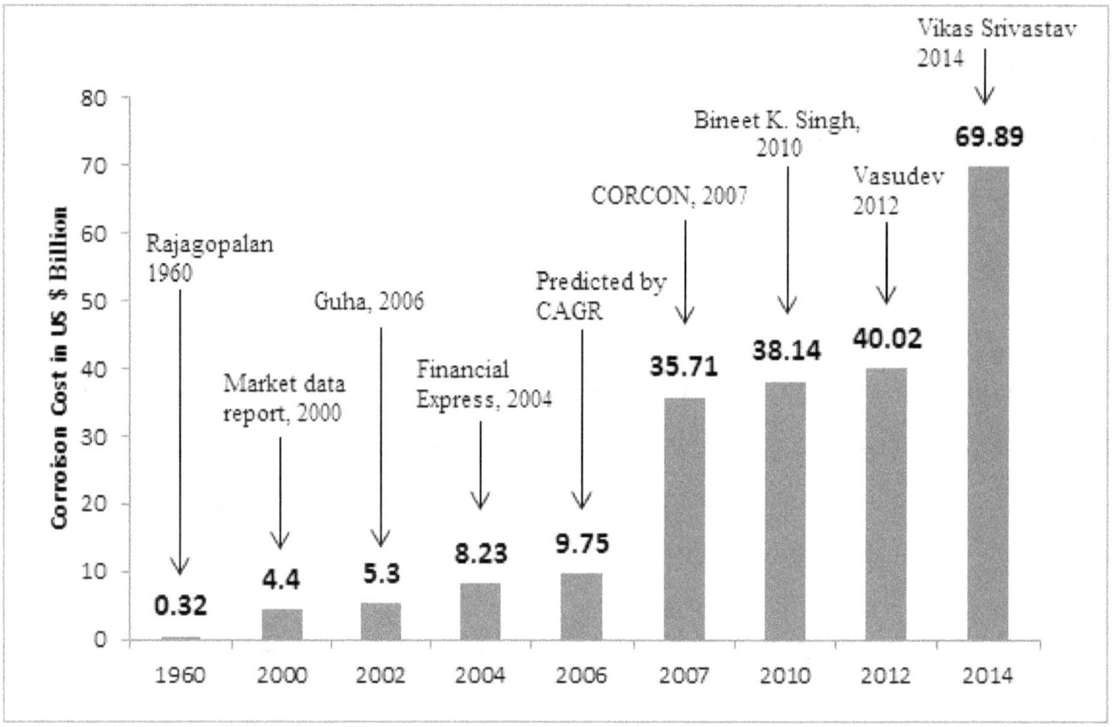

Figure 1.2 Sources of corrosion cost in India

Harvey Hack, President of NACE International, told Financial Chronicle, that one third of the cost (around \$10 billion to \$15 billion), can be saved by using available practices and technologies in our routine working environments. NACE is conducting India's first impact study of corrosion for reducing costs associated with it in numerous industries.

1.1.4 Definition of Corrosion

Corrosion involves the reaction of a metallic material with its environment and is a natural process in which the metal is trying to revert to the chemically combined state in which it is found in the earth's crust.

It is defined as ***"The deterioration of materials by chemical interaction with their environment".***

1.1.5 Chemistry of Corrosion

Corrosion reactions are electrochemical in nature and when a metal is immersed in a solution, positively charged metal ions tend to pass into the solution, leaving electrons (**Oxidation reaction**) i.e the accumulation of negative charge on the metal. This potential difference between the metal and solution is called the electrode potential which retard the dissolution of metal ions and further dissolution will continue only if the electrons released can pass to a site on the metal surface where a cathodic reaction is possible.

$$\underset{\text{Atoms on metal surface}}{M} \rightarrow \underset{\text{Ions in Solution}}{M^{n+}} + \underset{\text{Electrons in metal}}{n\,e^-} \qquad (1)$$

At a cathodic site the electrons react with some reducible component of the electrolyte and are themselves removed from the metal (**Reduction reaction**). In acid solutions, electrons can react with hydrogen ions, adsorbed on the metal surface from the solution, to produce hydrogen gas [6, 7]

$$\underset{\text{Adsorbed on metal surface}}{2H^+} + \underset{\text{In metal}}{2e^-} \rightarrow \underset{\text{Gas}}{H_2} \qquad (2)$$

Figure 1.3 represent the mechanism of corrosion and the corroding piece of metal is described as a “mixed electrode” since simultaneous anodic and cathodic reactions are proceeding on its surface.

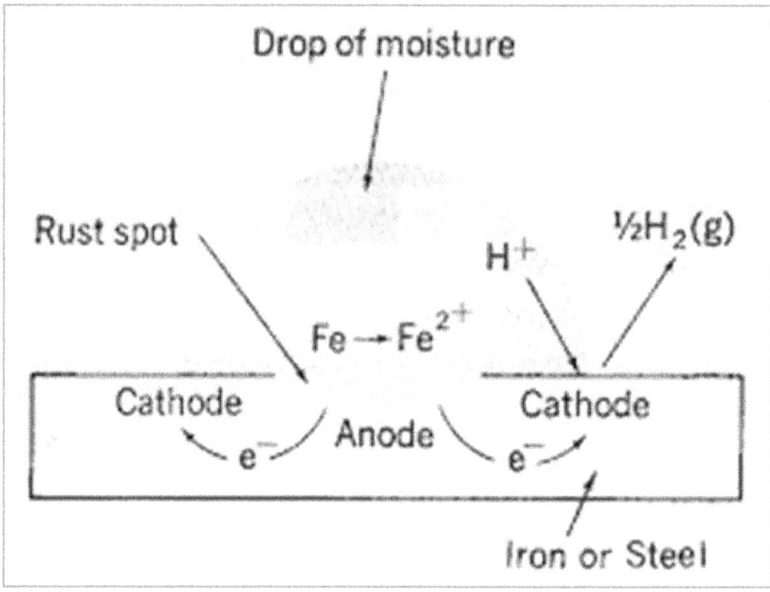

Figure 1.3 Mechanism of corrosion of iron

1.1.6 Factors Influencing Corrosion

The rate and extent of corrosion depend mainly on (i) The nature of the metal and (ii) The nature of the environment.

i. *Nature of the Metal*

a. **Position in the galvanic series:** The extent of corrosion depends upon the position of the metal in the galvanic series. Greater the oxidation potential, the greater is the rate of corrosion. When two metals are in electrical contact, the metal higher up in the galvanic series becomes anodic and suffers corrosion.

b. **Relative areas of the anode and cathode:** The rate of corrosion is more when the area of the cathode is larger as the demand for electrons will be more and this results in increased corrosion rate.

c. **Purity of the metal:** The impurities present in a metal create heterogeneity and thus galvanic cells are set up with distinct anodic and cathodic areas in the metal. The higher the percentage of impurity present in a metal, the faster is the rate of corrosion of the anodic metal.

d. **Physical state of the metal:** Metal components subjected to unevenly distributed 3stresses are easily corroded. Caustic embrittlement takes place in stressed parts such as bends, joints and rivets in boilers.

e. **Nature of the oxide film:** Metals such as Mg, Ca and Ba form oxides whose volume is less than the volume of the metal. Hence, the oxide film formed will be porous, through which oxygen diffuse and bring about further corrosion. On the other hand, metals like AI, Cr and Ni form oxides whose volume is greater than that of the metal and the nonporous oxide film so formed will protect the metal from further corrosion.

f. **Solubility of the products of corrosion:** If the corrosion product is soluble in the corroding medium, the corrosion of the metal will proceed faster.

ii. Nature of the Environment

a. **Temperature:** Corrosion increases with temperature. A passive metal may become active at a higher temperature.

b. **Effect of pH:** The possibility of corrosion with respect to pH of the solution and the electrode potential of the metal can be correlated with the help of a Pourbaix diagram.

c. **Humidity:** Atmospheric corrosion of iron is slow in dry air but increases rapidly in the presence of moisture as the moisture acts as the solvent for the oxygen in the air.

d. **Nature of the electrolyte:** If the conductance of electrolyte is more, the corrosion current is easily conducted and hence the rate of corrosion is increased.

e. **Concentration of oxygen and formation of oxygen concentration cells:** The corrosion rate increases with increasing supply of oxygen. The region where oxygen concentration is lesser becomes anodic and suffers corrosion.

1.1.7 Classification of Corrosion

Corrosion occurs in several widely differing forms. Classification is usually based on one of the three factors: ***Nature of the corrodent, mechanism of corrosion and appearance of the corroded metal.***

Almost all corrosion problems and failures encountered can be associated within anyone of the five general categories as mentioned below

1. **General / Uniform corrosion**

It occurs when a chemical or electrochemical attack occurs over a large area in a uniform manner.

- ***Atmospheric Corrosion*** – It is the corrosion or degradation of material exposed to the air and its pollutants rather than immersed in a liquid.
- ***Galvanic Corrosion*** – Galvanic corrosion occurs when a metal or alloy is electrically coupled to another metal or conducting non-metal in the same electrolyte.
- ***General biological Corrosion*** – It is the corrosion of metals over the entire exposed surface in aqueous environments.
- ***Motel salt Corrosion*** – It is the corrosion of metals due to molten or fused salts.
- ***Liquid metal Corrosion*** – It is the types of corrosion found in liquid metal / containment / component combinations.
- ***High temperature Corrosion*** – It is the corrosion by direct reaction of exposed metals to oxidizing agents at elevated temperatures like in oxidation, sulfidation, carburization, etc.

2. **Localized corrosion**

It is a type of corrosion where all or most of the metal loss occurs at discrete areas on the metal surface.

- ***Film form*** – It is the corrosion that occurs on metallic surfaces coated with thin organic film, typically 1 mm thick, characterized by the appearance of fine filaments in semi-random directions from one or more sources.
- ***Pitting Corrosion*** – A highly localized corrosion attack that results in holes is referred to as pitting. Pits may be isolated or localized and of virtually any configuration. They occur at defects or imperfections in a protective or passive film.
- ***Crevice Corrosion*** – Crevice corrosion is a highly localized attack occurring in a crevice or an otherwise shielded area when a material is exposed to a stagnant corrosive media.
- ***Localized Microbiological Corrosion*** – It is a case where biological organisms are the sole cause or an accelerating factor in the localized corrosion

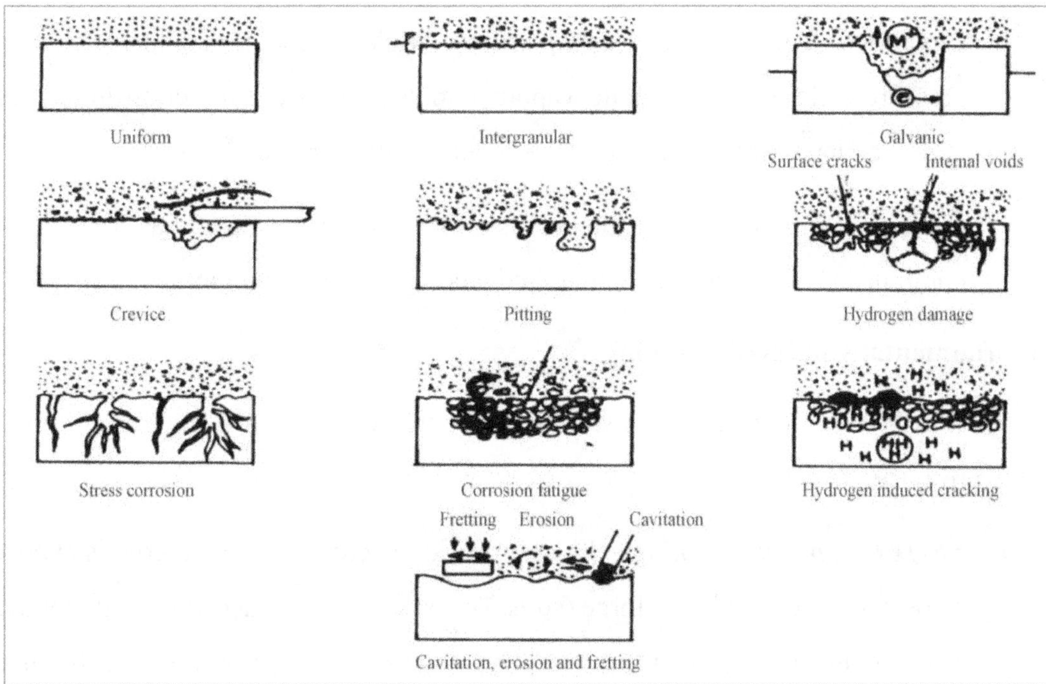

Figure 1.4 Forms of corrosion

3. Metallurgically influenced corrosion

In these types of corrosion attack metallurgy plays a significant role.

- ***Intergranular Corrosion*** - Intergranular corrosion occurs when the grain boundaries are attacked in preference to the material matrix.

- ***Dealloying (Selective Leaching)*** **-** Dealloying occurs when one of the elements is removed from the metal matrix, leaving an altered residual structure. It is commonly identifiable by a color change or a drastic change in mechanical strength.

4. Mechanically assisted degradation

It is a form of corrosion attack where velocity, abrasion, hydrodynamics etc. play a major role.

- ***Erosion*** **-** Erosion occurs when the velocity of the fluid is sufficient to remove protective films from the metal surface. It is often a combination of erosion (mechanical damage) with corrosion (electrochemical damage).
- ***Fretting Corrosion*** – Fretting occurs when motion between surfaces either removes protective films or mechanically removes material from surfaces in relative motion.
- ***Cavitation & Water Drop Impingement*** – Metal surface in contact with a liquid, pressure differentials generate vapour bubbles, upon encountering high-pressure zones, collapse and cause explosive shocks to the surface.
- ***Fatigue Corrosion*** – It is the corrosion that occurs in metals as a result of the combined action of a cyclic stress and a corrosive environment.

5. Environmentally induced cracking

It is the type where the corrosion occurs in the forms of cracking that are produced in the presence of stress.

- ***Stress Corrosion Cracking*** - Stress corrosion occurs when a material is exposed to a corrosive media while a force (stress) or pressure is applied. The material usually remains undamaged with the exception of cracks that grow through the material matrix.
- ***Liquid Metal Embrittlement*** – It is the brittle failure of a normally ductile metal when coated with a thin film of a liquid metal and subsequently stressed in tension.
- ***Solid Metal Embrittlement*** – It is similar ti liquid metal embrittlement and occurs below the melting point of the solid in certain liquid metal embrittlement couples.

1.1.8 Corrosion Prevention

The field of corrosion measurement, control and prevention covers a very broad spectrum of technical activities. All methods, measures and procedures aimed at the avoidance of corrosion damages are called ***corrosion prevention***. Corrosion prevention can be achieved in the following ways:

1. Conditioning the metal: This can be sub-divided into two main groups:

a. ***Coating the metal*** - Coating can be applied between metal and environment. The coating may consist of

i. Another metal, e.g. zinc or tin coatings on steel.

ii. A protective coating derived from the metal itself, e.g. aluminium oxide on "anodised" aluminium.

iii. Organic coatings, such as resins, plastics, paints, enamel, oils and greases.

b. ***Alloying the metal*** - To produce a more corrosion resistant alloy, e.g. stainless steel, in which ordinary steel is alloyed with chromium and nickel. Stainless steel is protected by an invisibly thin, naturally formed film of chromium sesquioxide Cr_2O_3.

2. Conditioning the corrosive environment – This can be achieved as follows:

a. ***Removal of oxygen*** - By the removal of oxygen from water systems in the pH range 6 - 8 one of the components required for corrosion would be absent. The removal of oxygen could be achieved by the use of strong reducing agents e.g. sulphite.

b. ***Corrosion inhibitors*** - A corrosion inhibitor is a chemical additive, when added to a corrosive aqueous environment reduces the rate of metal wastage.

3. Electrochemical control

Since corrosion is an electrochemical process, the rate of corrosion reactions may be controlled by passing anodic or cathodic currents into the metal. If electrons are passed into the metal and the anodic reaction will be stifled while the cathodic reaction rate increases. This process is called cathodic protection. In certain chemical environments it is sometimes possible to achieve anodic protection, i.e., passing a current which takes electrons out of the metal and raises its potential.

1.2 Introduction to Corrosion Inhibitors

Corrosion processes, can be controlled by compounds known as inhibitors which adsorb on the reacting metal surface. The term adsorption refers to molecules attached directly to the surface, normally only one molecular layer thick, and not penetrating into the bulk of the metal itself.

1.2.1 Types of Inhibitors

Inhibitors can be divided into two main categories (**Figure 1.5**).

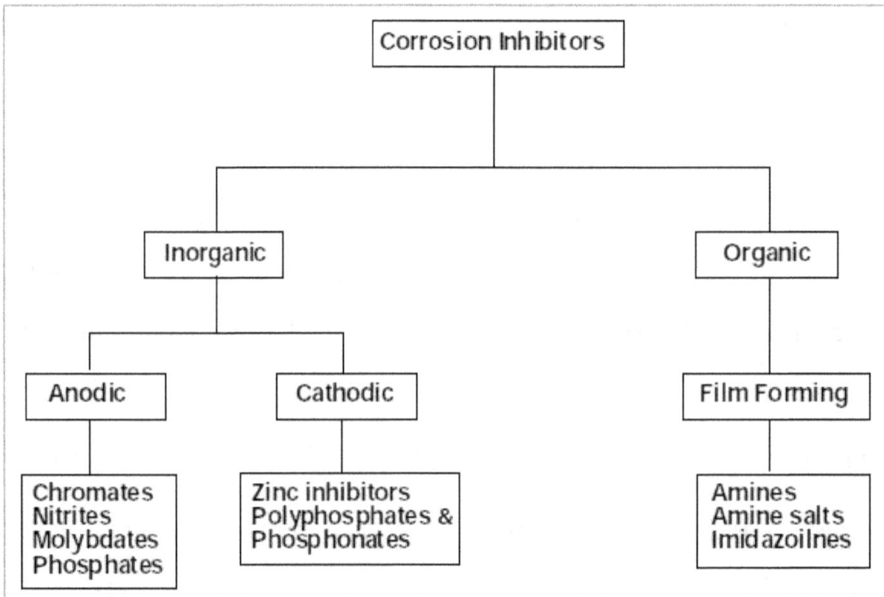

Figure 1.5 Types of Inhibitors

1. Inorganic inhibitors

Inorganic inhibitors are used mainly in boilers, cooling towers, and fractionation units whereas organic inhibitors are used mainly in oil field systems. Inorganic inhibitors are further divided into anodic and cathodic inhibitors.

i. ***Cathodic inhibitors*** **-** Cathodic inhibitors are chemical compounds which inhibit the cathodic reaction of the corrosion cell. Examples of cathodic inhibitors:

a. **Compounds** such as $ZnSO_4$, $MgSO_4$, $Ca(HCO_3)_2$ that form insoluble film that isolates the cathodic regions from the corrosive medium.

b. **Oxygen scavengers,** reacts with dissolved oxygen and remove it from the corrosive environment.

c. **Cathodic Poisons,** compounds such as arsenic (As), mercury (Hg), antimony (Sb) salts when added to acidic corrosive environments slowdown the cathodic H_2 evolution reaction.

ii. ***Anodic inhibitors:*** Anodic inhibitors are chemical compounds which inhibit the anodic reaction of the corrosive cells. Examples:

a. **Compounds** such as Na_2SiO_3, Na_2CO_3 and Na_3PO_4 form an insoluble film which isolates the anodic sites from the corrosive solution.

b. **Passivators,** these are oxidizing agents which react with the Fe^{2+} ions to form a passive oxide film on the anodic sites of the corrosion cells.

c. **Chromate inhibitors**, Chromate, or dichromate ions accept electrons and forms chromium (III) oxide which combines with iron oxide to form a mixed oxide. This mixed oxide forms a deposit that tightly attaches to the metal surface.

d. **Nitrite inhibitors,** perform similarly to chromates. They form a protective iron oxide film that passivates steel surfaces.

2. Organic inhibitors (Adsorption inhibitors)

These are organic compounds which contain a polar group that adsorb on the metallic surface and isolate it from the corrosive solution. Organic inhibitors typically affect both anodic and cathodic areas of corrosion cells. They lay down an organic film on the entire metal surface.

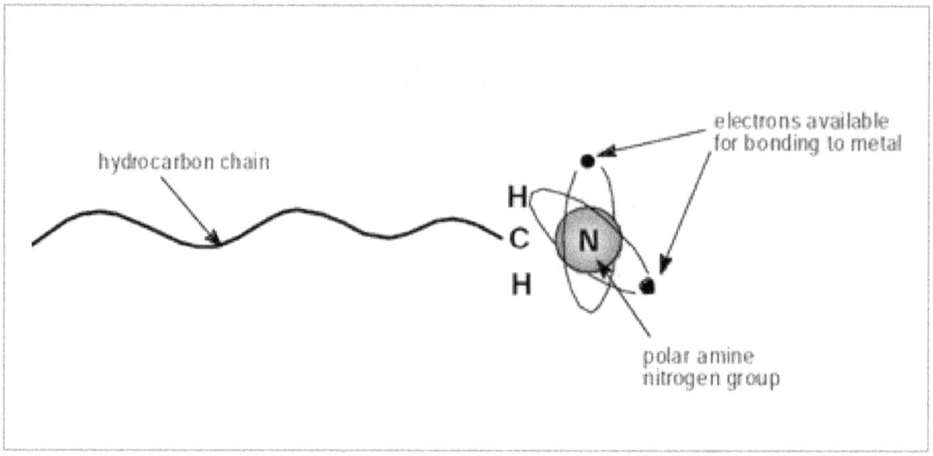

Figure 1.6 Organic inhibitor molecule

Their effectiveness depends upon the following conditions:

- The electrical potential of the metal
- The chemical structure of the inhibitor molecule
- The size and shape of the inhibitor molecule

An organic corrosion inhibitor molecule consists of a hydrocarbon chain that is attached to a strongly polar functional group. Polar functional groups are based on nitrogen, sulfur, or oxygen (**Figure 1.6**) and has a pair of unshared electrons. These electrons are available to bond with metal surfaces.

Organic inhibitor molecules attach to metal surfaces by chemisorption and physical adsorption. If an organic inhibitor is introduced into the system, the unshared electrons of the polar amine group form a chemisorption bond with the metal surface. Physical adsorption does not involve sharing of electrons with the metal surface and is weaker than chemisorption. **Figure 1.7** shows the molecular structures of some common oil field corrosion inhibitors.

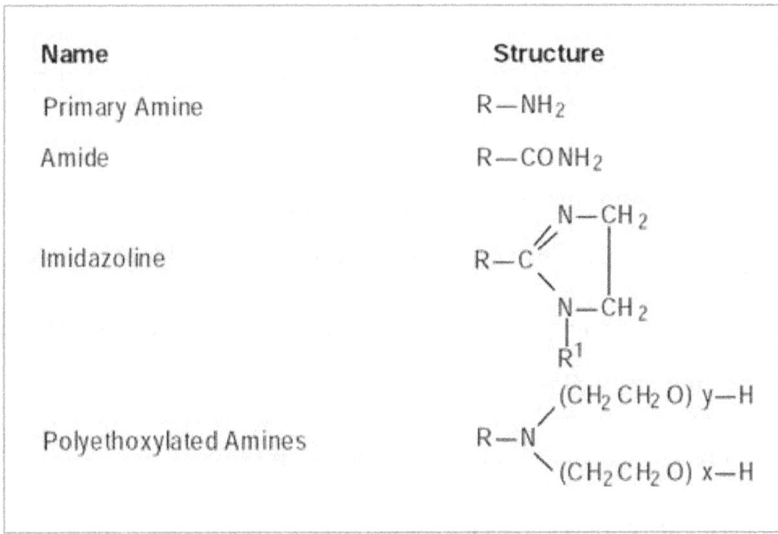

Figure 1.7 Molecular structure of common oil field inhibitors

1.2.2 Mechanism of Inhibition

A corrosion inhibitor may act in a number of ways:

i. It may restrict the rate of the anodic process or the cathodic process by simply blocking active sites on the metal surface.

ii. It may act by increasing the potential of the metal surface so that the metal enters the passivation region where a natural oxide film forms.

iii. A further mode of action of some inhibitors is that the inhibiting compound contributes to the formation of a thin layer on the surface which stifles the corrosion process.

1.3 Introduction to Corrosion Monitoring Techniques

Corrosion monitoring is one of the basic functions required for safe and efficient industrial operations.

1.3.1 Need for corrosion monitoring

The rate of corrosion indicates the working period of any process equipment during which it can be usefully and safely operated. Corrosion monitoring is the practice carried out to assess and predict the corrosion behavior and the objectives of corrosion monitoring are:

1. To provide information on the state of operational equipment with the intention of avoiding unplanned shut-downs, occurring due to unforeseen deterioration of the plant.
2. To provide information on the interrelation between corrosion processes and operating variables to allow more efficient use of the plant.
3. To provide information that plant inspection departments may use to prevent safety failures and potential disasters.
4. To assess levels of contamination of process fluids.

1.3.2 Corrosion Monitoring Techniques

Corrosion measurement is the quantitative method by which the effectiveness of corrosion control can be evaluated. A wide variety of corrosion measurement techniques exists, including:

Non Destructive Testing

- Ultrasonic testing
- Radiography & Thermography
- Intelligent pigs
- Eddy current/magnetic flux

Analytical Chemistry

- pH measurement
- Dissolved gas (O_2, CO_2, H_2S)
- Metal ion count (Fe^{2+}, Fe^{3+})
- Microbiological analysis

Operational Data

- pH
- Flow rate (velocity)
- Pressure
- Temperature

Fluid Electrochemistry

- Potential measurement
- Potentiostatic measurements
- Potentiodynamic measurements
- A.C. impedance

Some corrosion measurement techniques can be used on-line, constantly exposed to the process stream, while others provide off-line measurement, such as that determined in a laboratory analysis. Some techniques give a direct measure of metal loss or corrosion rate, while others are used to infer that a corrosive environment may exist.These techniques basically fall into seven categories.

i. CEION:

CEION is a new ratiometric metal loss measurement device with resolution 100 times better than ER devices. It is ideal for monitoring oil/gas production and processing systems, sub-sea, sub-surface applications and sand erosion.

ii. Electrical Resistance (ER)

ER probes provide a basic measurement of metal loss, at any time, as frequently as required. The ER technique measures the change in Ohmic resistance of a corroding metal. Corrosion on the surface of the element decrease its cross-sectional area which increases its electrical resistance. The increase in resistance is related to metal loss as a function of time which is the corrosion rate. ER probes have the following advantage:

- Direct corrosion rates can be obtained.
- Probe remains installed in-line until operational life has been exhausted.
- They respond quickly to corrosion upsets and can be used to trigger an alarm.

iii. Electrochemical Methods

Since corrosion is an electrochemical process, there exist a number of electrochemical methods for corrosion monitoring. The essential difference between ER and Electrochemical techniques is that ER measurements provide information on total loss of material, whereas electrochemical techniques give rate information.

a. **Linear Polarization Resistance (LPR):** In LPR technique, a small voltage (or polarization potential) is applied to an electrode in solution. The current needed to maintain a specific voltage shift (typically 10 mV) is directly related to the corrosion on the surface of the electrode in the solution. By measuring the current, a corrosion rate can be derived. The advantage of the LPR technique is that the measurement of corrosion rate is made instantaneously.

b. **Galvanic Monitoring** - The galvanic monitoring technique, also known as Zero Resistance Ammetry (ZRA) is another electrochemical measuring technique. With

ZRA probes, two electrodes of dissimilar metals are immersed in solution and a natural voltage (potential) difference exits between the electrodes. The current generated due to this potential difference relates to the rate of corrosion.

c. **Electrical Impedance Spectroscopy (EIS):** In order to access the charge transfer resistance or polarization resistance in proportional to the corrosion rate at the monitored interface, EIS uses a model of the interface. EIS monitors the electric response of the metal/environment interface to the applied AC signal over a frequency spectrum usually in the range of 10 kHz - 50 μHz. Most important is the solution resistance which can be separated from the actual polarization resistance. EIS determine the properties of the surface layers such as pore resistance and film capacitance.

d. **Electrochemical Noise (EN)** - Electrochemical noise (EN) records the fluctuations in the corrosion potential and current. Current noise uses current variations between two similar electrodes whereas potential noise is based on variations in potential between a working electrode and a stable reference electrode. Qualitatively, EN data has been utilized to identify localized corrosion.

iv. Weight Loss Coupons

The Weight Loss technique is the best known and simplest of all corrosion monitoring techniques. The method involves exposing a specimen of material, the coupon (**Figure 1.8**) to a process environment for a given duration, then removing the specimen for analysis. The basic measurement which is determined from corrosion coupons is weight loss; the weight loss taking place over the period of exposure being expressed as corrosion rate.

Figure 1.8 Metal coupons used for weight loss studies

Advantages of weight loss coupons are that:

- The technique is applicable to all environments - gases, liquids, solids/particulate flow.
- Visual inspection can be undertaken.
- Corrosion deposits can be observed and analyzed.
- Weight loss can be readily determined and corrosion rate easily calculated.
- Localized corrosion can be identified and measured.
- Inhibitor performance can be easily assessed.

v. Tafel Polarization

The Tafel extrapolation method for determining corrosion rate was used by Wagner and Traud to verify the mixed potential theory. Corrosion is an electrochemical process in which, the anodic reaction is the ionization of the metal atoms leaving their electrons and in the cathodic reaction, and the free electrons within the metal are taken up by chemical species such as O_2 and H_2O in reduction reactions. Under standard conditions, the potential of the corrosion cell is the difference between the cathodic and the anodic half-cell potential which generates a current flow.

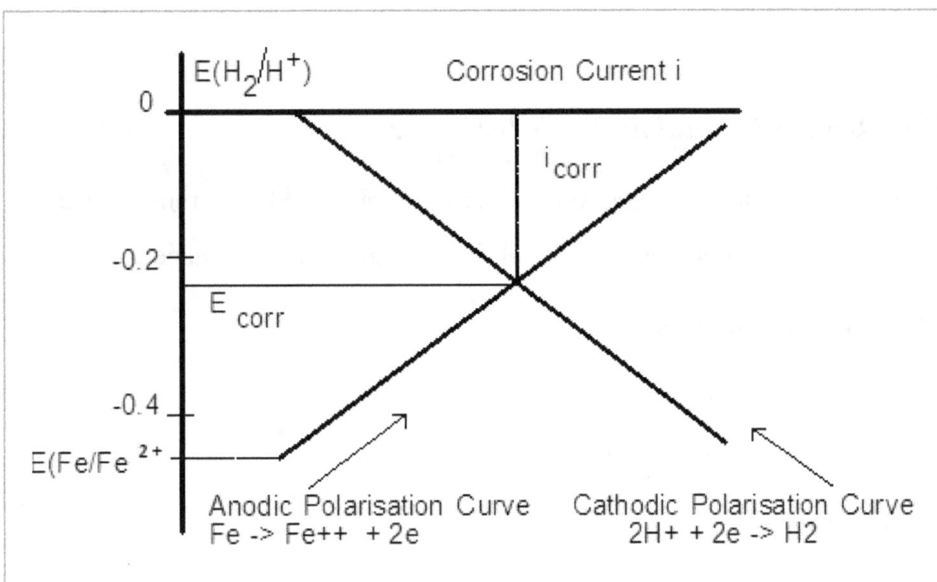

Figure 1.9 Typical polarization curve for iron

Polarization is the effect of reducing the potential difference between the half cells. The Tafel polarization technique method involves the measurement of over potential for various current densities and a plot of η vs log I (**Figure 1.9**) is made. The slopes give Tafel constants (ba & bc) and the intercept corresponds to $I_{corr.}$ At the point of intersection of the two polarisation curves, a metal will be freely corroding. The potential at which this occurs is called the free corrosion potential (E_{corr}) which generates corrosion current (I_{corr}). From the corrosion current, it is possible to calculate corrosion rates.

vi. Hydrogen Penetration Monitoring

In acidic process environments, hydrogen is a by-product of the corrosion reaction. Hydrogen generated in such a reaction can be absorbed by steel particularly when traces of sulphide or cyanide are present. This may lead to hydrogen induced failure. The concept of hydrogen probes is to detect the amount of hydrogen permeating through the steel by mechanical or electrochemical measurement.

vii. Specialized Techniques

Biological Monitoring - Biological monitoring and analysis identify Sulphate Reducing Bacteria - SRB's which consume sulphate from the process stream and generate sulphuric acid, a corrosive which attacks production plant materials.

Sand / Erosion Monitoring - These are devices which are designed to measure erosion in a flowing system.

viii. Non-Destructive Testing (NDT) Techniques

These include ultrasonics, radiography, thermography, eddy current measurement and various others. Such techniques usually fall into the area of plant inspection.

ix. Analytical Techniques

Areas of interest under this heading are iron counts, chloride counts, oxygen, conductivity and pH measurements, flow measurements and temperature.

Drug as an inhibitor

The organic acid inhibitor that contains oxygen, nitrogen and/or sulphur is adsorbed on the metallic surface blocking the active corrosion sites [8-9]. Although the most effective [10-11] and efficient organic inhibitors are compounds that have π-bonds, it present biological toxicity and environmental harmful characteristics [12-13]

Most of the commercial and organic inhibitors are toxic in nature; therefore replacement by environmentally benign inhibitors is necessary few non-toxic compounds have been investigated as corrosion inhibitors by some researchers. Modified plastic waste was used as cheap and safe corrosion inhibitors for metals and alloys in different aqueous media. The use of pharmaceutical compounds offer interesting possibilities for corrosion inhibition due to the presence of hetero atoms like nitrogen, sulphur and oxygen in their structure and are of particular interest because of their safe use, high solubility in water and high molecular size. Some of the azosulpha and antimalarial drugs have been reported as good corrosion inhibitors, the use of piperazine derivative drug as a corrosion inhibitor has been reported.

Coatings for corrosion protection

In order to give corrosion protection, the protective coatings should provide

- Resistance to penetration of ions of salts [14]
- Resistance to action of osmosis
- Expand and contract with the underlying surface
- Long life

1.4. Paints:

Coating is defined as a material which is applied on to a surface and appears as a continuous film after drying properties of the coating is defined by the type of binders, pigments and miscellaneous additives used in the formation. They are anti-corrosion coating material. They keep the surface of the metals out of contact with air and moisture [15-17].

Constituent of paint

- Binders or resin
- Pigment

- Solvent
- Extenders
- Driers

a) Resin

This is a clear varnish or gummy base which attaches itself to the surface of the substance to be coated and solidifies there. Some vehicles dry by simply by evaporation of solvent but most harden by some chemical action such as polymerization. Vehicles are classified into two, nonconvertible material and convertible coatings.

(i) Nonconvertible material

It simply dries after evaporation. Main examples are non-drying alkyl resins, shellac, manila gum, bitumen, waxes, cellulose, chlorinated rubber and vinyl and acrylic resin.

(ii) Convertible coating

Once deposited their nature change under the action of atmospheric oxygen or by action of catalyst or heat. In this molecules join up to form large and tough polymeric films. Ex: various drying oils like linseed oil, Tung oil, and china wood oil etc. Resins used in surface coating are

- Epoxide
- Rosin
- Alkyd
- Coumarone
- Urea formaldehyde

Most surface coating use a blend of synthetic resin in order to obtain best result. Examples of synthetic resin varnishes are alkyd resin, alkyd styrene copolymer resin, amino resin, silicones etc.

b) Binders

The binder is a polymer, a film forming material, which can be either organic or inorganic in nature. At present binders are predominantly synthetic. Some of the binders are listed below.

- Natural resin which when heated gets polymerized.
- Alkyds – polymerization products of polyhydric alcohols with polycarboxylic acids in combination with drying oils.
- Amino resin – urea formaldehyde, melamine formaldehyde.
- Phenolic – condensation binders synthesized from phenol or formaldehyde in acidic and basic media.
- Epoxies – single pack epoxy ester and double pack epoxy / polyamide and epoxy / amine system.
- Vinyl – polymer of vinyl chloride, vinyl acetate.
- Silicone resin- silicon combined with various organic compounds.

Different types of binders

i. Epoxy binders

Epoxy power coatings are the oldest and one of the largest classes of thermosetting power coatings. The most commonly used cross-liners are dicyandiamide or a modified dicyandiamide. 2-methyllimidazole is widely used a catalyst. Epoxy power coatings have good mechanical properties, adhesion, and corrosion protection. Their applications include pipes, electrical equipment's and primers and so on.

ii. Acrylic binders

Hydroxyl-functional acrylics can be cross-linked with blocked isocyanates or glycolurils and carboxylic acids. The epoxy-functional acrylics are made with glycidyl methacrylate as a co-monomer and are cross-linked with di carboxylic acid resin. They have superior detergent resistance, and are used as washing machines.

iii. Silicone binders

The silicone and silicone / polyester resin are used in high heat resistance power coatings. The silicone power coating requires low processing temperatures then other power coatings [18-19].

c) Pigments

They are solid materials which are suspended in the form of small discrete particles in the medium and which served to give the paint color, opacity, increased abrasion resistance, certain chemical properties and improved body.

Characteristics of pigments:

- Color
- Hiding power
- Particle size and shape
- Ease of wetting
- Oil absorption
- Bulking value
- Corrosion inhibitor
- Resistance of light, heat, water and chemicals.

Pigment volume concentration (PVC)

PVC is defined as the percentage of pigment volume in total volume solids of the paint. The PVC is simply the ratio of the volume of pigment to the total volume of solids pigment plus binder. PVC like volume solids is always expressed as a percentage and hence the 100 in the equation.

$$PVC = \frac{\text{Volume of pigment}}{\text{Volume of pigment} + \text{volume of binder}} \times 100$$

Maximum gloss is obtained when PVC is very low. PVC also affects other properties such as film flexibility, adhesion, impact, resistance, moisture permeability etc..,

Solvents: This is used to render the pigment / binder mixture sufficiently fluid for application to a surface in the form of a uniform film. The thinner should then evaporate completely to leave s film consisting of pigment and binder only. Thinner in common use includes a wide range of organic solvents.

In addition to major ingredients paint may also contain driers, anti-skinning agents, antifoaming agents, anti-setting agents; fungicides etc. are used to assist dispersion of the pigments.

The main purpose is to provide the surface coating material in a form in which it can readily de applied by brushing, spraying or dipping. They evaporate after surface coating material has been applied. It varies in nature from to organic material such as ester, hydro carbons and ketone. Solvents can be classified in 3 broad categories

Weak hydrogen bonding solvents

Eg, toluene, xylene, high flash atomic naphtha's

Hydrogen- bonded acceptors solvents

Eg – methyl ethyl ketone, acetone, methyl isobutyl ketone

Hydrogen bond donor- acceptor solvents

Eg – methyl alcohol, isopropyl alcohol, n- butyl alcohol

Extenders

They are added in order to increase the body of paint and thus to improve the abrasion other mechanical resistance.

Driers

They are catalysts added to paint in order to increase the rate of setting and to improve the final strength.

Types of coating

a) Primer

A primer is the first coat of a system. Its principal function is to obtain adequate adhesion and afford good protection to the substrate. The addition is impaired by the presence of oil grease and for this reason, it is essential that the surface to be painted should be properly cleaned. It is desirable to incorporate inhibitive pigment in the primer, so as to minimize corrosion.

b) Intermediate coat

Intermediate coats are typically used in coating systems, which are designed for precise purpose. The primary purpose of an intermediate coat generally has rather high pigment to vehicle ratio, so that it is a flat coat with superior bond.

c) Top coat

The choice of top coat is very important because it provide the following function such as

- Providing a resistant seal for the coating system
- Forming the primary wall to the situation
- Providing a hard and war resistant exterior
- Providing a nick look

The top coat is denser then intermediate coat since it is formulated with a lower pigment to vehicle ratio. The paint is generally applied by brush spray and some special methods such as, electro coating, roller coating and fluidized bed methods.

1.5 Introduction to Computational Chemistry

Computational Chemistry stimulates chemical structures and reactions based on the fundamental laws of physics. It allows chemists to study chemical phenomena by running calculations on computers. Some methods can be used to model not only stable molecules, but also short – lived, unstable intermediates and even transition states [20-21]. Over the last two decades molecular modelling are capable of accurately predicting structures, energetics, reactivity and other properties of molecules. Computational chemistry plays an important role in enabling chemists to: (i) explain and rationalize known chemistry and (ii) explore new or unknown chemistry.

1.5.1 Advantages of Computational study

- Computational calculations are easy to perform, whereas experiments are often difficult.
- Computational calculations are becoming less costly, whereas experiments are becoming more expensive.
- Computational calculations can be performed on any system, even those that don't exist, whereas many experiments are limited to relatively stable molecules.
- Computational calculations are safe, whereas many experiments have an intrinsic danger associated with them.

1.5.2 Overview of Computational Chemistry

There are two broad area within computational chemistry related to the structure of molecules and their reactivity: *Molecular Mechanics and Electronic Structure Theory [22].*

1. Molecular mechanics – Molecular mechanics stimulations use law of classical physics to predict the structures and properties of molecules, which is characterized by a particular ***force field.*** A ***force field*** has three components:

i. A set of equations defining hoe the potential energy of a molecule varies with the locations of its atoms.

ii. A series of atom type defining the characteristics of an element within a specific chemical content.

iii. One or more parameter set the fit the equation set and atom type to the experimental data.

Limitations of Molecular mechanics

- No force field can be used generally for all the molecular systems.
- Neglect of electrons means molecular mechanics cannot be used for chemical problems where electronic effect predominates and molecular properties which

depend on the electronic details are also not reproducible by molecular mechanics methods.

2. Electronic structure methods **-** Electronic structure methods use the laws of quantum mechanics rather than classical physics as their basis. Quantum mechanics states that the energy and other related properties of a molecule are obtained by solving the time-independent Schrödinger wave equation [23]. Electronic structure method is characterized by their various mathematical approximations to its solution. There are three classes of electronic structure methods:

i. *Semi – empirical methods,* such as AM1, MINDO/3 and PM3, use parameters from experimental data to simplify the computation.

ii. *Ab initio methods,* means "from the beginning" or "from first principles" in which their computations are based on the laws of quantum.

iii. *Density Functional methods,* include the effect of electron correlation – the electrons in a molecular system react to one another's motion and attempt to keep out of one another's way.

1.5.3 Theory of quantum chemistry

The theoretical foundation for computational chemistry is the time-independent Schrödinger wave equation:

$$\check{H}\Psi = E\Psi \qquad (3)$$

Ψ is the wave function. $\check{H}$ is the Hamiltonian operator and is associated with the observable energy. E is the total energy of the system [24]

The Hamiltonian, $\check{H}$, is an operator. It contains all the terms that contribute to the energy of a system:

$$\hat{H} = \hat{T} + \hat{V} \qquad (4)$$

Which describes a system consisting of nuclei and electrons. Born–Oppenheimer approximation, is used to arrive at the solution to the above equation.

1.5.4 Model Chemistry

The Schrödinger wave equation can be approximated much more closely for small systems than for large ones. However, using different levels of theory for different size molecules makes comparing results among systems unreliable [25-26]. The implementation of such a theoretical model is termed as **'Theoretical Model Chemistry'** or **'Model Chemistry'**. The other desirable features of model chemistry include:

- *Size consistency* – the results given for a system of molecules separated from one another must be equal to the sum of the results obtained foe each individual system.
- Reproducing the *exact solution for the relevant n-electron problem* – method should yield same results as the Schrödinger wave equation to the greatest extent possible.
- *Variational* – the energies predicted must be bound to the real energy resulting from the solution of the Schrödinger wave equation.
- *Efficient* – calculation with a method ought to be practical with existing computer technology.
- *Accurate* – a method should produce highly accurate quantitative results.

Model chemistries are characterized by the combination of theoretical procedure and basis set.

Basis Set – A *Basis Set* is a mathematical representation of the molecular orbitals within a molecule. The basis set can be interpreted as restricting each electron to a particular region of space. Larger basis set impose fewer constraints on electrons and more accurately approximate exact molecular orbitals.

Methods – There exist a hierarchy of procedures corresponding to different approximation methods (commonly referred to as different levels of theory) as listed below.

Keyword	**Method**	**Availability**
HF	Hartree - Fock self-consistent field	Through 2nd derivatives
B3LYP	Beck-style 3-parameter Density Functionl Theory (using the Lee-Yang-Parr correlation functional)	Through 2nd derivatives
MP2	2nd order Moller –Plesset Perturbation Theory	Through 2nd derivatives
MP4	2nd order Moller –Plesset Perturbation Theory (including singlet, doublet, triplet and qudrauples by default)	Energies only
QCISD(T)	Quadratic CI (single, doubles & triples)	Energies only

Open vs Closed Shell – A third component for every calculation involves how the electron spin is handled using an *Open shell or a Closed shell* model (*unrestricted* and *restricted* calculation).

- For closed shell molecules, having an even number of electrons divided into opposite spin, a spin restricted model is default. It uses doubly occupied orbitals, each containing two electrons of opposite spin. Restricted closed shell calculations force each electron pair into a single spatial orbital.

- Open shell system with unequal numbers of spin up and spin down electrons are usually modelled by a spin unrestricted model. Open shell calculations use separate spatial orbitals for the spin up and spin down electrons (α & β respectively).

  Unrestricted calculations are required for systems with unpaired electrons, including: Molecules with odd number of electrons (eg. Ions), excited states, systems with unusual electronic structure (eg. 2 or more unpaired outer electrons), processes such as bond dissociation.

References

[1] U. R. Evans, “An Introduction to Metallic Corrosion”, 2nd Ed. London: Arnold, 1972.

[2] H. H. Uhlig, “Corrosion and Corrosion Control”, 3rd Ed. New York: John Wiley and Sons, 1985.

[3] M. G. Fontana & N. D. Greene, “Corrosion Engineering”, 3rd Ed. New York: McGraw-Hill Book Company, 1986.

[4] N. Hackerman, “A view of the History of Corrosion and its Control”, In: Gundry, R. D. Ed. Corrosion 93, Planery and Keynote Lectures, Texas: NACE, Houston, 1– 5, 1993.

[5] Bineet Kumar Singh, Munmun Basak, Martin West, Indranil Guha, “Estimating the Cost of Corrosion in Indian industry”, Petrotech-2010, New Delhi, India, and Paper ID: 20100815.1, 2010

[6] Raj Narayan, “An Introduction to Metallic Corrosion and its Prevention”, Mohan Primlani for Oxford & IBH Publishing Company, 1983.

[7] V. S. Sastri, “Corrosion Inhibitors: Principles and Applications”,copyrighted to John – Wiley & Sons Ltd, England, 2001

[8] Verma C, Olasunkanmi LO, Ebenso EE, Quraishi MA, Obot IB ,"Heterocyclic compounds as corrosion inhibitors for mild steel in hydrochloric acid medium—correlation between electronic structure and inhibition efficiency”, *J Phys Chem* C vol.120, pp.11598–11611, 2016.

[9] Verma C, Olasunkanmi O.I.B, Ebenso E.E, Quraishi M.A, 2,4-Diamino-5-(phenylthio)-5H-chromeno [2,3-b] pyridine-3-carbonitriles as green and effective corrosion inhibitors: gravimetric, electrochemical, surface morphology and theoretical studies. *RSC Adv,* vol.6, pp.1–50, 2016.

[10] Bendaif H, Melhaoui A, Azzouzi. M El, Legssyer B, Hamat T, Elyoussfi A, Aouniti AE, Ouadi Y, Aziz M,"Eco-friendly pancratium foetidum pom extracts as corrosion inhibitors for mild steel in 1M HCl media”. *J Mater Environ Sci*, vol.7, pp.1276–1287, 2016.

[11] Bentiss F, Lagrenée M ,"Heterocyclic compounds as corrosion inhibitors for mild steel in hydrochloric acid medium—correlation between electronic structure and inhibition efficiency". *J Mater Environ Sci*, vol.2, pp.13–17, 2011

[12] Gece G," Drugs: a review of promising novel corrosion inhibitors". *Corros Sci,* vol 53, pp.3873–3898, 2011

[13]. Quraishi, M. A, Chauhan, D. S, Saji, V. S., "Heterocyclic Organic Corrosion Inhibitors: Principles and Applications", *Elsevier Inc*, Amsterdam, 2020.

[14] Zhaoxia Li,,Wenbin Yang,Qiangliang Yu,Yang Wu,Daoai Wang,Jun Liang,, and Feng Zhou, "New Method for the Corrosion Resistance of AZ31 Mg Alloy with a Porous Micro-Arc Oxidation Membrane as an Ionic Corrosion Inhibitor Container", *ACS publications*, vol.35, no.5, pp.1134–1145, 2019

[15] Saji, V.S., & Thomas, *J. Nanomaterials for corrosion control,* Current Science, vol.92, no. 1, 2007

[16] Harris, C., & Mc Lachlan, R. Clark, Colin-Micro reform-impacts on films: aluminium case study. Melbourne: Industry commission,1998

[17] Mahindru, D.V. Ms. Priyanka Mahendru, protective Treatment of alumunium its alloys. *Global Journal of Research in Engineering Version* 1.0, vol.11, no.3, 2011

[18] Huang, Y. "Electrochemical Evaluation of advanced anodized aluminium and Chromate-Free conversion Coatings", A Dissertation presented to the faculty of the graduate school University of Southern California, 2009

[19] Environmental protection agency. CFR part 61: National emission standard for hazardous air pollutants. National Emission standard for radon emissions from phosphogypsumstacks, federal register, vol.64, no.2, pp.5574-5580, 1990

[20] B. K. Sen, "Quantum Chemistry Including Molecular Spectroscopy", Tata McGraw- Hill, Publishing Company Ltd, New Delhi, 1996.

[21] A. K. Chandra, "Introductory Quantum Chemistry", Tata McGraw Hill Publishing Company Limited, New Delhi, 1998.

[22] A. S. Negi, S. C. Anand, “A Text Book of Physical Chemistry”, New Age International (P) Limited, New Delhi, 2002.

[23] James B. Frishman and A. Frisel, “Exploring Chemistry with Electronic Structure Methods”, Gaussian, Inc. Pittsburgh, PA.

[24] J. B. Foresman, “Ab Initio Techniques in Chemistry: Interpretation and Visualization”, Chapter 14 in *'What Every Chemist Should Know about Computing'*, Ed. M. L. Swift and T.J. Zielinski, ACS Books, Washington, D.C., 1996.

[25] A. Szabo and N. S. Ostlund, “Modern Quantum Chemistry”, McGraw-Hill, New York, 1982.

[26] W. J. Hehre, L. Radom, P.V.R. Schleyer and J. A. Pople, “Ab Initio Molecular Orbital theory, Wiley, New York, 1986.

Chapter II

CHAPTER II

REVIEW OF LITERATURE

2.1 Mild Steel

1. **Meng *et al.,*** studied that two novel pyridine Schiff base derivatives, namely 3-pyridinecarboxaldehyde-4-phenyl thiosemicarbazide (3-PCPTC) and 4-pyridinecarboxaldehyde- 4-phenylthiosemicarbazide (4-PCPTC), were synthesized and their corrosion inhibition properties on mild steel in 1.0 M HCl were studied through experimental and theoretical measurements. Weight loss, electrochemical impedance spectroscopy and potentiodynamic polarization results indicate that both compounds exhibit a good inhibition effect and that 3-PCPTC shows better inhibition efficiency than 4-PCPTC. Meanwhile, the adsorption process obeys the Langmuir adsorption isotherm model, and thermodynamic parameters were calculated and discussed. Scanning electron microscopy and X-ray photoelectron spectroscopy analyses confirm the adsorption and protective abilities of the two studied inhibitors on mild steel in HCl. UV-vis spectroscopic studies show that the corrosion inhibitors interact with mild steel in HCl solution to form Fe-inhibitor complexes. Furthermore, quantum chemical calculations and molecular dynamics simulations were performed. According to the experimental and theoretical results, the inhibition mechanism was deduced.

2. **Arukalam *et al.*,** studied that the corrosion inhibition of mild steel in 1.0 M H_2SO_4 solution by ethyl hydroxyethyl cellulose has been studied in relation to the concentration of the additive using weight loss measurement, EIS, polarization, and quantum chemical calculation techniques. The results indicate that EHEC inhibited corrosion reaction in the acid medium and inhibition efficiency increased with EHEC concentration. Further increase in inhibition efficiency is observed in the presence of iodide ions, due to synergistic effect. Impedance results reveal that EHEC is adsorbed on the corroding metal surface. Adsorption followed a modified Langmuir isotherm, with very high negative values of the free energy of adsorption (ΔGads). The polarization data indicate that the inhibitor was of mixed type, with predominant effect on the cathodic partial reaction. The frontier molecular orbitals, HOMO (the highest occupied molecular orbital) and LUMO (the lowest unoccupied molecular orbital) as well as local reactivity of

the EHEC molecule, were analyzed theoretically using the density functional theory to explain the adsorption characteristics at a molecular level. The theoretical predictions showed good agreement with experimental results.

3. Dakeshwar Kumar Verma, reported that Bryophyllum pinnatum leaves extract has been evaluated as mild steel corrosion inhibitor in 0.5 M HCl solution by gravimetric analysis and Scanning electron microscope techniques. Effect of temperature and inhibitor concentrations on the corrosion behaviour of mild steel were studied and found that inhibition efficiency increases as inhibitor concentration increases but decreases as the temperature increases. The activation energy and thermodynamic parameters were studied, results direct that corrosion process involve the exothermic process and adsorption of inhibitor molecules follows the physical adsorption on metal surface. Linearity of Langmuir adsorption isotherm shows the formation of monolayer over the metal surface by deposition of inhibitor molecules. SEM analysis reveal the smooth surface which indicate the protective adsorption of inhibitors molecules on metal surface. The results reveal that the extract of bryophyllum pinnatum act as an effective corrosion inhibitor for the mild steel corrosion in 0.5 M HCl medium.

4. Palaniappana *et al.,* investigated that the unused expired clinical drugs are a major environmental hazardous pollutant due to the pharmaceutically active organic compounds present in it. Until recently, a quantum of unused drugs is being disposed through sewers, drainage, sediment is a threat to the public health. Several researchers have already reported the potential use of expired drugs as corrosion inhibitors in both acidic and alkaline medium. In this paper, we have reported three different drug molecules which have crossed the expiry date but not beyond six months. The drugs considered are rabeprazolesodium, domperidone and benfotiamine. Mild steel (MS) specimens were immersed 5 days with and without drug molecule in 3.5 wt% medium film formed on MS surface. These films were then analysed for crystallinity, surface morphology, iron complex, DFT and functional groups by XRD, SEM, and 3D profilometer, FTIR, UV and EPR. The experimental and theoretical results are good agreement with each other

5. Khamis *et al.*, determined that the synergistic inhibitory action of cetylpyridinium chloride (CPC) and different halides on the corrosion of mild steel in 0.5 M H2SO4 has been investigated using electrochemical methods and scanning electron microscopy (SEM). Experimental results showed that the protection efficiency (Picor) of CPC improved in the presence of the different halides with different extents. Chemisorption was proposed from the trend of Picor and the values of DG0 ads. The synergism parameter is found to be greater than unity indicating that the enhanced Picor caused by the addition of the halides to the CPC is due to a co-operative adsorption of both species.

6. Jamiu K. Odusote *et al.*, found that the inhibition of corrosion of mild steel in sulphuric acid by acidic extract of Jatropha Curcas leaves has been studied using weight loss and thermometric measurements. It was found that the leaves extract act as a good corrosion inhibitor for mild steel in all concentrations of the extract. The inhibition action depends on the concentration of the Jatropha Curcas leaves extract in the acid solution. Results for weight loss and thermometric measurements indicate that inhibition efficiency increase with increasing inhibitor concentration. The adsorption of Jatropha curcas leaves extract on the surface of the mild steel specimens obeys Langmuir adsorption isotherm. Based on the results, Jatropha curcas leaves extract is recommended for use in industries as a replacement for toxic chemical inhibitors.

7. M. Faiz *et al.*, Corrosion inhibition effect of the crude extracts (hexane, dichloromethane, methanol) from the bark of Cryptocarya nigra and three alkaloids named N-methylisococlaurine 1, N-methyllaurotetanine 2 and atherosperminine 3 isolated from the Cryptocarya nigra dichloromethane extract (CNDE) were investigated for mild steel corrosion in 1 M HCl solution. An electrochemical impedance study showed that CNDE and 2 reduced the corrosion significantly through a charge transfer mechanism with inhibition efficiency of 91.05% and 88.05%, respectively. Potentiodynamic polarization data indicated that CNDE acted through anodic type inhibition while 2 was a mixed type inhibitor with predominant anodic effectiveness. DGads values calculated from the Langmuir adsorption isotherm plots for CNDE (-28.2 kJ mol^{-1}) and 2 (-13.2 kJ mol^{-1}) suggested that they adsorbed on the mild steel surface via a physisorption mechanism. Scanning electron microscopy micrographs and elemental composition studies confirmed the formation of a protective film over the metal surface. Wastewater quality parameters of all the inhibitors demonstrated good biodegradability as their values were within the permissible limits to discharge for irrigation and horticultural uses.

8. Fathabadia *et al.*, reported that Tolyltriazole (TTA) is a well-defined corrosion inhibitor for copper and copper alloys. However, there is little literature about its corrosion inhibition performance for mild steels in corrosive environments. This paper studied the electrochemical behaviour of TTA in 0.5 M HCl solutions. Also, the morphology and nature of TTA layers on the steel surface were investigated. Electrochemical results showed that TTA is an excellent corrosion inhibitor for mild steel in acidic media with an efficiency of 91% for 0.07 M concentration. The results also indicated that TTA is a mixed-type inhibitor. XRD analysis revealed that the inhibition mechanism of TTA is based on the formation of an organic film due to the physically adsorbed molecules of TTA. AFM and EDS results showed that the formed layers decrease the adsorptivity of corrosive elements (Cl-) on the steel surface. Density Functional Theory (DFT) study confirmed the experimental results.

9. G. Karthik *et al.*, reported that the inhibition performance of atenolol on mild steel in 1 M hydrochloric acid solution was studied by weight loss and electrochemical methods. The results show the inhibition efficiency was found to increase with increasing the concentration of the inhibitor from 50 to 300 ppm. The maximum inhibition efficiency 93.8% was observed in the presence of 300 ppm inhibitor (in case of potentiodynamic polarization). The inhibition action of atenolol was explained in terms of adsorption on the mild steel surface. The adsorption process follows Langmuir isotherm via physical adsorption. Electrochemical Impedance spectroscopic technique (EIS) exhibits one capacitive loop indicating that, the corrosion reaction is controlled by charge transfer process. Polarization measurements showed that the inhibitor is of a mixed type. The results obtained from the different methods are in good agreement. The surface morphologies of mild steel were examined by Fourier-transform infrared (FT-IR) spectroscopy, scanning electron microscope (SEM). Further, the computational calculations are performed to find a relation between their electronic and structural properties.

10. Ahmed A. Farag ***et al .,*** reported that the recovery shrimp waste protein (RSWP) was extracted, characterized and tested their inhibition effects as a corrosion inhibitor for carbon steel in 1 M HCl solution using different electrochemical techniques. The results have been shown that the corrosion rate decreases with increasing inhibitor concentration. To verify the morphology of RSWP film formed on the carbon steel surface, scanning electron microscopy (SEM) was used. The inhibition mechanism of the extracted RSWP is discussed in terms of adsorption on the carbon steel surface. The results of the energy adsorption were showed chemisorption effect for on the carbon steel surface.

11. Gunavathy ***et al.,*** investigated that the inhibition effect of unripe fruit peel extract of Musa acuminata (Cultivar variety – Nendran) (MNP) on corrosion of mild steel in 1 N HCl has been investigated by weight loss and electrochemical impedance spectroscopy (EIS) with various concentrations of the extract. The effect of temperature on the corrosion inhibition of mild steel in the temperature range of 30 °C – 80 °C was carried out. The results indicate that MNP extract act as an effective inhibitor in the acid environment and is of mixed type inhibitor having efficiency as high as 96% at 2% inhibitor concentration. The inhibition efficiency of MNP extract increases with the increase of concentration but decreases with the increase in temperature. The inhibitor achieves its inhibition by physical adsorption of nutrients of the peel extract on the surface of the mild steel. The experimental data revealed that the adsorption occurred according to the Langmuir and Temkin adsorption isotherm.

12. T. H. Ibrahim ***et al.,*** reported that the inhibition effect of Fig leaves extract on corrosion of mild steel in 2M acid solution was investigated using traditional weight loss measurements and various electrochemical techniques. Corrosion inhibition efficiency of Fig leaves extracts was evaluated using weight loss measurements for effect of various solution concentrations. In addition, electrochemical techniques were performed to evaluate both concentration and temperature changes of test solution. Nyquist plots showed that the higher the Fig leaves extract concentration, the higher the charge transfer resistance and the lower the value of the double layer capacitance. Both weight loss and electrochemical results showed similar trend indicating that Fig leaves extracts adsorption on mild steel followed the Langmuir adsorption isotherm at all test temperatures. The change in free energy and the activation energy values indicate that the adsorption is physical. Inhibition efficiency of as high as 87% were achieved for Fig leaves extract of above 200 ppm.

13. M. Manickam ***et al*** **.,** found that the ability of gum exudates of Azadirachta indica (GAI) to inhibit corrosion on mild steel in 1 mol L−1HCl has been studied using mass loss, polarization, and impedence measurements. The effect of temperature (303–323 K) and immersion time of 1, 2, 4, 6, and 12 hours on corrosion behavior of mild steel was examined. Gum exudates decrease the corrosion rate up to a concentration of 80 ppm and at 323 K temperature. GAI adsorb chemically onto the surface of the mild steel while it obeys Langmuir adsorption isotherms. Polarization studies show GAI as mixed mode inhibitor. Surface studies ascertain that a shielding layer was formed on the mild steel surface.

14. Alfred I. Onen ***et al.,*** were studied about the behaviour of titanium (IV) oxide, TO as a corrosion inhibitor of aluminium and mild steel in H_2SO_4 solutions. It has been investigated using the "Absorbance difference" technique at 303K and 313K. The percentage inhibition efficiencies (%IE) increased with concentration of TO and decrease with increasing temperature in the order: 59.62 % MS < 73.34 % Al. The activation energy, Ea of aluminium (Al) and mild steel (MS) corrosion increased with increasing concentration. The experimental results fit the Langmuir adsorption isotherm. Physisorption mechanism has been proposed for the inhibitor based on the Ea values (Ea _ 9.82 kJmol-1 for Al and Ea _7.18 kJmol-1for MS). The negative values of heat of adsorption, Qads obtained in this study implies that the adsorption of TO molecules on the Al and MS surfaces is spontaneous.

15. Rupesh Kushwah and R. K. Pathak was state that the inhibition of mild steel corrosion in 0.5 M sulphuric acid solution by Aspirin drug as an eco-friendly and commercially available inhibitor was studied at different temperature by weight loss technique. It was found that the test drug has a promising inhibitory action against corrosion of mild steel in the medium investigated. The inhibition efficiency was found to increase with a corresponding increase in the concentration of the inhibitor. It was also found that the adsorption as well as the inhibition process followed first-order kinetics and Aspirin adsorbs on the steel surface according Langmuir isotherm. Adsorption enthalpy were determined and discussed. Effect of temperature was also investigated and activation parameters were evaluated.

16. W. Niouri *et al* ., were investigated that the corrosion inhibition of mild steel in molar HCl by two newly synthesized benzodiazepine derivatives, namely: 10-benzyl-pyrrolo[2,1-*c*][1,4]benzodiazepine-5,11-dione (BZD=2O), and 10-benzyl-pyrrolo[2,1*c*][1,4] benzodiazepine-5,11-dithione (BZD=2S); at 308 K using electrochemical and weight loss measurements. The results showed that these compounds were good inhibitors and the inhibiting efficiency increased with rise concentration of the studied inhibitors. Compound BZD=2S showed better protection properties even at higher temperatures. BZD=2S was adsorbed on the mild steel surface according to a Langmuir isotherm adsorption model. The associated kinetic and thermodynamic parameters for both metal dissolution and inhibition processes have been determined and discussed.

17. M. A. Quraishi *et al.*, were investigated the corrosion inhibition properties of Ketorol on mild steel corrosion in 1.0 M HCl solution using electrochemical impedance spectroscopy (EIS), potentiodynamic polarization and weight loss methods. EIS data were interpreted using CPE in equivalent circuit. Adsorption of the inhibitor on the mild steel surface followed the Langmuir adsorption isotherm. The value of free energy of adsorption ($\Delta G°$ads) is -35.52 kJ mol-1 indicated that adsorption of Ketorol is a spontaneous process. Ketorol exhibited the highest inhibition efficiency 92.36 % at concentration 300 ppm.

18. Sutiana Junaedi *et al.,* was synthesized 1, 5 -Dimethyl-4-((2-methylbenzylidene) amino)-2-phenyl-1*H*-pyrazol-3(2*H*)-one DMPO), it to be evaluated as a corrosion inhibitor. The corrosion inhibitory effects of DMPO on mild steel in 1.0 M HCl were investigated using electrochemical impedance spectroscopy (EIS), potentiodynamic polarization, open circuit potential (OCP) and electrochemical frequency modulation (EFM). The results showed that DMPO inhibited mild steel corrosion in acid solution and indicated that the inhibition efficiency increased with increasing inhibitor concentration. Changes in the impedance parameters suggested an adsorption of DMPO onto the mild steel surface, leading to the formation of protective films. The novel synthesized corrosion inhibitor was characterized using UV-Vis, FT-IR and NMR spectral analyses. Electronic properties such as highest occupied molecular orbital energy, lowest unoccupied molecular orbital energy (E_{HOMO} and E_{LUMO}, respectively) and dipole moment (μ) were calculated and discussed. The results showed that the corrosion inhibition efficiency increased with an increase in the E_{HOMO} values but with a decrease in the E_{LUMO} value.

19. N. O. Eddy *et al.,* were studied the inhibitory and adsorption potentials of 2-(4-methoxybenzylideneamino) acetic acid (MBDAC), 2-hydroxy-3-methoxybenzaldehyde (HMBA) and 4-methoxybenzaldehyde (MBA) using experimental (gravimetric and gasometric) and computational chemistry simulation approaches. The results obtained from the experimental approach revealed that the trend for the inhibitory potentials of the three anisoles is HMBA > MBDAC > MBA. The observed trend is attributed to the structural differences resulting from the introduction of –OH, =O and NCH_2CO_2H substituents into the respective compounds. The adsorption of the three inhibitors on mild steel surface is spontaneous and supported the mechanism of chemical adsorption. Computational chemistry simulations was carried out by comparing calculated quantum chemical parameters (the energy of the frontier molecular orbitals, the energy gap, dipole moment and logP) of the studied anisoles in gas and aqueous phases with experimentally obtained % inhibition efficiencies. The tests gave excellent correlations for PM6, PM3, AM1, RM1 and MNDO Hamiltonians. Correlations between experimental and predicted inhibition efficiencies were also excellent. Local selectivity study using the Fukui function and global softness indices indicated that the likely sites for electrophilic and nucleophilic attacks are in the enol bonds of the inhibitors.

20. The corrosion inhibition of mild steel in 0.5 M H_2SO_4 by ethanolic extract from the leaves of *Chromolaena odorata L.* (LECO) has bccn studicd as a possible source of green inhibitor using the gravimetric technique in the temperature range 303-333 K, which were studied by **I.B. Obot *et al*.**, Results obtained showed that the LECO functioned as a corrosion inhibitor for mild steel in the acidic environment. Inhibition efficiency increased with extract concentration but decreased with temperature. At a concentration as low as 5 %v/v of the extract, the percentage inhibition reached about 95 % at 303 K and 89 % at 333 K. The thermodynamic parameters for the adsorption of this inhibitor on the metal surface such as were calculated using the Langmuir adsorption isotherm. The activation (*Ea*) energy and the heat of adsorption (*Qads*) of the corrosion inhibition process suggest that the LECO molecules are physically adsorbed on the metal surface. The ethanolic

Chlomolaena odorata L. leaves extract was found to be an excellent potential corrosion inhibitor because of the presence of essential oils, steroids, triterpenes, as well as the high content of a series of complex flavonoids, chalcones and flavones in the extract.

2.2 Aluminium

21. Mohammed A. Amin *et al.,* was study the influence of three selected polyacrylic acids (PAAs) with different molecular weights (PAA1 = 1800, PAA2 = 11,000 and PAA3 = 14,000 g mol_1) on the corrosion inhibition of Al in weakly alkaline solutions (pH 8 and 10) at 30 _C. Measurements were conducted under different experimental conditions using chemical (weight loss) and electrochemical (potentiodynamic polarization and impedance) techniques, complemented with ex situ energy dispersive X-ray (EDX) examinations of the electrode surface. Electrochemical frequency modulation (EFM), a non-destructive corrosion measurement technique that can directly give values of corrosion current without prior knowledge of Tafel constants, is also presented here. The results demonstrated that these polymers inhibit the alkaline corrosion of Al. The inhibition effect of these polymers is due to their adsorption on Al surface. The isoelectric point (IEP) of aluminium oxide (pH 9) seems to be an important factor controlling corrosion inhibition and adsorption of the three polymers. The three polymers inhibit the corrosion reaction of aluminium excellently at pH 8, but less effectively at pH 10. Polarization measurements showed that the three polymers act as mixed-type inhibitors. The inhibition efficiencies of these polymers increase with increasing concentration, molecular weight and immersion time. Results obtained from the chemical and electrochemical measurements are in good agreements.

22. Umoren *et al.*, were studied the corrosion inhibitive effects of *Aningeria robusta* extract for aluminium in 2 M HCl solution and the influence of potassium iodide additives on the inhibition efficiency was assessed using hydrogen evolution method at 30 and 60 °C. Inhibition efficiency was determined by comparing the corrosion rates in the absence and presence of additives. The trend of inhibition efficiency with temperature was used to propose the mechanism of inhibition. It was found that the *Aningeria robusta* extract acts as an inhibitor for acid-induced corrosion of aluminium. Inhibition efficiency (%I) of the extract increased with an increase in concentration of the *Aningeria robusta* extract and with increase in temperature. On the other hand, inhibition efficiency (%I) synergistically increased on addition of potassium iodide but decreased with increase in temperature. Inhibitor adsorption characteristics were approximated by Langmuir adsorption isotherm at all the concentrations and temperatures studied. The mechanism of chemical adsorption is proposed for the *Aningeria robusta* extract and physical adsorption for the extract-iodide mixture from the trend of inhibition efficiency with temperature and from the calculated values of activation energy and heat of adsorption.

23. I.B. Obot and N.O. Obi-Egbedi were report for the first time the inhibitive action of the root of ginseng on aluminium corrosion in HCl solution using weight loss method at 30-60°C. Results obtained showed that ginseng root functioned as an effective and excellent inhibitor in the acid medium. Corrosion rate increased both in the absence and presence of inhibitor with increase in temperature. Corrosion rate was also found to decrease in the presence of inhibitor compared to the free acid solution. Inhibition efficiency increases with increase in concentration of the inhibitor but decreases with increase in temperature reaching a maximum of 93.1% at 30°C at 50 % v/v concentration of ginseng. Addition of iodide ions to the root extracts of ginseng enhances the inhibition efficiency considerably and the effect is more pronounced at higher temperatures. The adsorption of extract components onto the aluminium surface was found to be a spontaneous process and to follow the Freundlich adsorption isotherm. The free energies, enthalpy and entropy for the adsorption process as well as the energy of activation, enthalpy of activation and entropy of activation for the dissolution process were determined and discussed. A mechanism of physical adsorption of the root components on the surface of the metal is proposed for the inhibition behaviour.

24. The role of some surfactants in the corrosion of Al in 1 M HCl has been investigated using weight loss and galvanostatic polarization techniques it is studied by **G.Y. Elewady *et al.*,** Results showed that the inhibition occurs through adsorption of the inhibitor molecules on the metal surface. The inhibition efficiency was found to increase with increasing inhibitor concentration and decreased with increasing temperature which is due to the fact, that the rate of corrosion of Al is higher than the rate of adsorption. The adsorption of the compounds on the metal surface is found to obey Freundlich adsorption isotherm. The inhibiting actions of those compounds are considerably enhanced by the addition of KI, due to the increase of the surface coverage and therefore indicate the joint adsorption of these compounds and iodide ions. Thermodynamic parameters for adsorption and activation processes were determined. Galvanostatic polarization data indicated that these compounds act as mixed-type inhibitors .Results obtained from the two techniques are in good agreement.

25. D.G. Ladha, U.J. Naik and N.K. Shah were studied the inhibitive effect of cumin extract as corrosion inhibitor for pure Aluminium in 1N HCl has been investigated by using Weight loss, Galvanostatic polarization and Electrochemical impedance spectroscopy (EIS) techniques. The results obtained reveal that inhibition efficiency increases with increase in concentration of inhibitors but decreases with increase in temperature. The value of apparent activation energy shows that the cumin extract acts as a good inhibitor for pure Aluminium in acid medium. The values of free energy of adsorption and heat of adsorption show that the adsorption of cumin extract on Aluminium obeys mixed adsorption and exothermic in nature. Thermodynamic consideration shows that adsorption of cumin extract follows Langmuir Adsorption isotherm. The Galvanostatic polarization measurement indicates that the inhibitor is of mixed type. Electrochemical impedance study results are in good agreement with weight loss and Galvanostatic polarization studies.

2.3 Quantum chemical studies

26.Mahendra Yadav *et al.*, investigated that three benzimidazole derivatives, namely, 4-(phenyl)-5-[(2-methyl-1*H*-benzimidazol-1-yl)methyl]-4*H*-1,2,4-triazole-3-thiol (Inh I), 4-(4-methylphenyl)-5-[(2-methyl-1*H*-benzimidazol-1-yl)methyl]-4*H*-1,2,4-triazole-3 -thiol (Inh II), and 4-(4-methoxyphenyl)-5-[(2-methyl-1*H*-benzimidazol-1-yl)methyl]-4*H*-1,2,4-triazole-3-thiol (Inh III), were synthesized and investigated as inhibitors for mild steel corrosion in 15% HCl solution using the weight loss, electrochemical polarization, and electrochemical impedance spectroscopy (EIS) techniques. It was found that the inhibition efficiency of these inhibitors increased with increasing concentration. The effect of temperature on the corrosion rate was investigated, and some thermodynamic parameters were calculated. Polarization studies showed that all three studied inhibitors were of mixed type in nature. The adsorption of the inhibitors on the mild steel surface in acid solution was found to obey the Langmuir adsorption isotherm. Scanning electron microscopy (SEM) was performed on inhibited and uninhibited mild steel samples to characterize the surface. The semiempirical AM1 method was employed for theoretical calculations, and the obtained results were found to be consistent with the experimental findings.

27. Lukman O. Olasunkanmi *et al.,* reported that the inhibition of mild steel corrosion in 1 M HCl by some quinoxalin-6-yl derivatives namely 1-[3-phenyl-5-quinoxalin-6-yl-4,5-dihydropyrazol-1-yl]butan-1-one (PQDPB), 1-(3-phenyl-5-(quinoxalin-6-yl) -4,5-dihydro-1H-pyrazol-1-yl)propan-1-one (PQDPP) and 2-phenyl-1-[3-phenyl-5-(quinoxalin-6-yl)-4,5-dihydropyrazol-1-yl]ethanone (PPQDPE) has been investigated using electrochemical studies and quantum chemical calculations. The results showed that PQDPP is the best corrosion inhibitor among the three compounds studied and the inhibition efficiency increases with increase in concentration for all the inhibitors. The adsorption of inhibitor molecules on mild steel surface was found to be spontaneous and obeyed the Frumkin adsorption isotherm. Scanning electron microscopy (SEM) images confirmed the formation of protective films of the inhibitors on mild steel surface. Quantum chemical calculations showed that the inhibitors have the tendency to be protonated in the acid and the results agree with experimental observations. Monte Carlo simulations were applied to search for the most stable configuration and adsorption energy for the interaction of inhibitors on Fe (110)/100 H2O interface. The results of the Monte Carlo simulations accord with the experimentally determined inhibition efficiencies. Different carbonyl substituents on the common nucleus of the three compounds obviously contributed to the difference in inhibition efficiency.

28. Priyanka Singh *et al.,* investigated that the corrosion inhibition efficiencies of three novel naphthyridines namely, 5-amino-9-hydroxy-2-phenylchromeno[4,3,2-de][1,6]-naphthyridine-4-carbonitrile (N-1), 5-amino-9-hydroxy-2-(p-tolyl)chromeno-[4,3,2-de][1,6] naphthyridine -4-carbonitrile (N-2), and 5-amino-9-hydroxy-2-(4-methoxyphenyl) chromeno [4,3,2-de][1,6]naphthyridine-4-carbonitrile (N-3) have been investigated for mild steel in 1 M HCl solution by using weight loss, electrochemical impedance spectroscopy, and potentiodynamic polarization methods. All three compounds show high inhibition activities at 6.54 × 10−5 M: N-1, 94.28%; N-2, 96.66%; and N-3, 98.09%. Electrochemical impedance spectroscopy analysis reveals an increase in polarization resistance due to the adsorbed inhibitor molecules on metal surface. Potentiodynamic polarization analysis reveals that all three compounds act as mixed-type inhibitors but of predominantly cathodic type. The adsorption of the studied compounds on mild steel surface follows the Langmuir adsorption isotherm. Surface morphology examined by using scanning electron microscopy and atomic force microscopy analysis shows a smoother surface for mild steel in the presence of naphthyridines in acidic solution. Quantum chemical parameters correlate well with the experimental results, which support higher inhibition efficiencies of N-3 and N-2 due to the electron-donating effects of −OCH3and −CH3 substituents, respectively, than of N-1, which is devoid of substituents. The magnitudes of the adsorption energies obtained from Monte Carlo simulations also agree with the trend of the experimental inhibition efficiency.

29. Jiyaul Haque *et al.*, reported that the corrosion inhibition property of ethyl 1,2,6-triphenyl-4-(phenylamino)-1,2,5,6- tetrahydropyridine-3-carboxylate (THP-1) and ethyl 2,6-bis(4-methoxyphenyl)-1-phenyl-4-(phenylamino)-1,2,5,6- tetrahydropyridine-3-carboxylate (THP-2) on mild steel (MS) in 1 M HCl have been studied using weight loss, electrochemical, surface morphology (SEM, AFM) and quantum chemical (QC) calculations methods. Results showed that both the functionalized tetrahydropyridines act as an excellent inhibition for mild steel (MS) and it exhibited the maximum inhibition efficiency (IE) of 90.67 % (THP-1) and 94.30% (THP-2) at 7.95×10-5 M concentration. Results further showed that adsorption of THPs on metallic surface obeyed the Langmuir adsorption isotherm. Potentiodynamic polarization (PDP) study revealed that studied compounds acted as mixed type inhibitors. Electrochemical impedance spectroscopy (EIS) study suggests that the investigated inhibitors inhibit corrosion by adsorbing at the metal/ electrolyte interfaces. The SEM, EDX and AFM studies showed that the THPs form a protective covering over the metallic surface that isolates the metal from aggressive acid solution and thereby protect form corrosive dissolution. Quantum chemical (QC) calculations carried out using DFT methods for neutral, solvated and protonated forms of inhibitors provides reasonable support to the experimental findings.

30. Hong Ju *et al.*, investigated Quantum chemical calculations based on DFT method were performed on three nitrogen-bearing heterocyclic compounds used as corrosion inhibitors for the mild steel in acid media to determine the relationship between the molecular structure of inhibitors and inhibition efficiency. The structural parameters, such as energy and distribution of highest occupied molecular orbital (HOMO) and lowest unoccupied molecular orbital (LUMO), the charge distribution of the studied inhibitors, the absolute electronegativity (χ) values, and the fraction of electrons (ΔN) transfer from inhibitors to mild steel were also calculated and correlated with inhibition efficiencies. The results showed that the inhibition efficiency of inhibitors increased with the increase in energy of HOMO and decrease in energy gap of frontier molecular orbital, and the areas containing N and O atoms are most possible sites for bonding the steel surface by donating electrons to the mild steel.

31.Iman Danaee *et al.*, investigated that the inhibitory effect of Oxethazaine drug, 2,2'-(2-Hydroxyethylimino)bis[N-(alphaalpha-dimethylphenethyl)-N-ethylacetamide] on corrosion of mild steel in 1M HCl solution was studied by weight loss measurements, electrochemical impedance spectroscopy and potentiodynamic polarization methods. The results of gravimetric and electrochemical methods demonstrated that the inhibition efficiency increased with an increase in inhibitor concentration in 1M HCl solution. The results from electrochemical impedance spectroscopy proved that the inhibition action of this drug was due to adsorption on the metal surface. Potentiodynamic polarization studies revealed that the molecule was a mixed type inhibitor. The adsorption of the molecule on the metal surface was found to obey Langmuir Adsorption isotherm. Potential of zero charge at the metal-solution interface was measured to provide the inhibition mechanism. The temperature dependence of the corrosion rate was also studied in the temperature range from 30 to 50 °C. Quantum chemical calculations were applied to correlate electronic structure parameters of the drug with its inhibition performance. The obtained theoretical results have been adapted with the experimental data.

32. Kalaiselvi Kathirvel *et al.*, found that the corrosion inhibition properties of 2,6-diphenylpiperidin-4-one (DPP) (1A) and 2,6-diphenyldihydro-2H-thiopyran-4(3H)-one (DPDT) (1B) for mild steel in 1 M phosphoric acid were studied using weight loss, potentiodynamic polarization and electrochemical impedance spectroscopic techniques.

The effect of temperature on the corrosion behavior of mild steel has been examined in the temperature range 303 - 328 K. The inhibition efficiency increases with increasing inhibitor concentration but decreases with increasing temperature. Potentiodynamic polarization studies indicated the mixed nature of inhibitors. The adsorption of the inhibitors on mild steel surface obeyed the Langmuir adsorption isotherms. The density functional theory (DFT) at the B3LYP/6- 31G (d) basis set level was performed on 1A and 1B to investigate the correlation between molecular structure and the corresponding inhibition efficiency (%). The quantum chemical parameters such as E_{HOMO}, E_{LUMO}, the energy gap ($\triangle E$), hardness (η), softness (S), dipole moment (μ), electron affinity (A), ionization potential (I), the absolute electronegativity (χ), the fraction of electron transferred ($\triangle N$), electrophilicity index (ω), the back-donation ($\triangle E_{Back\text{-}donation}$) and Mulliken population analysis have been calculated.

33. Eno E. Ebenso *et al.,* suggested that three thio semicarbazides, namely 2-(2-aminophenyl)-N phenyl hydrazine carbothioamide (AP4PT), N,2-diphenylhydrazinecarbothioamide (D4PT) and 2-(2-hydroxyphenyl)-N-phenyl hydrazine carbothioamide (HP4PT), were investigated as corrosion inhibitors for mild steel in H2SO4 solution using gravimetric and gasometric methods. The results revealed that they all inhibit corrosion and their % inhibition efficiencies (%IE) follow the order: AP4PT > HP4PT > D4PT. The %IE obtained from the gravimetric and gasometric experiments were in good agreement. The thermodynamic parameters obtained support a physical adsorption mechanism and the adsorption followed the Langmuir adsorption isotherm. Some quantum chemical parameters were calculated using different methods and correlated with the experimental %IE. Quantitative structure activity relationship (QSAR) approach was used on a composite index of some quantum chemical parameters to characterize the inhibition performance of the studied molecules. The results showed that the %IE were closely related to some of the quantum chemical parameters, but with varying degrees. The calculated/ theoretical %IE of the molecules were found to be close to their experimental %IE. The local reactivity has been studied through the Fukui and condensed softness indices in order to predict both the reactive centers and to know the possible sites of nucleophilic and electrophilic attacks.

34. F. El-Taib Heakal ***et al*****.,** investigated that the present study evaluates the effect of Camellia sinensis leaves (CSL) extract on the corrosion inhibition of carbon steel in produced water using potentiodynamic polarization, electrochemical impedance spectroscopy (EIS) and computational methods. The chromatographic investigation of the extract was performed by HPLC. It was found that CSL extract serves as a mixed-type corrosion inhibitor and its inhibitive performance increases with concentration. The corrosion inhibition was assumed to occur via adsorption of the primary constituents of CSL extract. Surface analysis (SEM and FTIR) was also carried out to establish the mechanism of corrosion inhibition. The outcomes obtained from the computational study were found to confirm experimental data.

35. Ahmed Al-Amiery ***et al*****.**, suggested that the corrosion inhibition of Schiff base, namely 2-((2-hydroxy-5-methoxybenzylidene) amino) pyridine (HMAP), for mild steel (MS) in a 1 M hydrochloric acid environment was investigated by means of weight loss and scanning electron microscopy techniques. Quantum chemical calculation based on density functional theory (DFT) was carried out on HMAP. Results illustrated that HMAP is a superior inhibitor for the corrosion of MS in 1.0M hydrochloric acid environment, and inhibition efficiency is higher than 90.0% at 0.5 g/L HMAP. Inhibition performance increases with regard to concentration increase and inhibition performance decreases when raising temperature. Adsorption of the inhibitor on the MS surface followed Langmuir adsorption isotherm and the value of the free energy of adsorption; Gads indicated that the adsorption of HMAP was a physisorption/chemisorption process. The DFT refers to perfect correlation with methodological inhibition performance.

36. Quy Huong Dinh ***et al.*****,** 1-Benzyl-3-phenyl-2-thiourea (BPTU) was studied as a steel corrosion inhibitor in 1.0 M HCl solution. Experimental methods were conducted including potentiodynamic polarization measurement (PPM), electrochemical impedance spectroscopy (EIS), and scanning electron microscopy (SEM) analysis. Quantum calculations were performed at B3LYP/6-311G(d,p). Hexamethylenetetramine (URO) was selected for comparison with BPTU results showed that BPTU with the concentration of 2 ×10−4 M and at the temperature of 30°C could protect the steel surface with the highest inhibition efficiency of 94.99% and 94.30% according to EIS and PPM, respectively. High temperature decreased BPTU's ability to inhibit the steel corrosion. The adsorption of BPTU on the steel surface is followed by the modified Langmuir isotherm. Quantum chemical calculations showed that the thiourea functional group is the main adsorption center of BPTU experimental results are completely consistent with theoretical calculations.

37. M. Ouakki *et al.,* found that the adsorption and corrosion inhibition properties of two imidazole derivatives namely, 2-(4-chlorophenyl)-1,4,5-triphenyl-1H-imidazole (IM-Cl) and 1,4,5-triphenyl-2-(p-tolyl)-1H-imidazole (IM-CH3) for mild steel in 0.5 M H 2SO4 solution are studied by electrochemical and computational calculations. The results obtained from the electrochemical methods show that IM-Cl and IM-CH3imparted high resistance and behave as mixed type inhibitors. Inhibition efficiency (IE %) increases with the increase of inhibitors concentration to attain 96 % and 91% at 103 M of IM-Cl and IM-CH3 respectively. EIS data is analyzed to model the inhibition process through appropriate equivalent circuit model. Thermodynamic and kinetic parameters controlling the adsorption process are calculated and discussed. DFT calculations are carried out at the B3LYP levels of theory with 6-31G (d,p) basis stein gas and aqueous phase for neutral and protonated forms. Quantum chemical calculations section of the study provides enough calculation and discussion on the relationship between corrosion inhibition and global reactivity descriptors.

38. Ahmed Y. Musa *et al.,* investigated that the inhibition ability of phthalhydrazide (PTD) for mild steel in 1 M HCl at 30 °C was investigated by electrochemical measurements [impedance spectroscopy (EIS) and potentiodynamic polarization techniques] and quantum chemical calculations. The frontier molecular orbital energy EHOMO (highest occupied molecular orbital), ELUMO (lowest unoccupied molecular orbital), and the Mulliken charge distribution were calculated and are discussed. Results showed that the inhibition efficiency of PTD increased with inhibitor concentration. The maximum corrosion inhibition efficiency was 77.6% at 2 mM PTD. Adsorption of the inhibitor followed the Langmuir adsorption isotherm. Adsorption of inhibitor molecules on mild steel surface occurred spontaneously and chemically. Quantum chemical calculations showed that the low performance of PTD as a corrosion inhibitor is due to the large energy gap (EHOMO - ELUMO).

39. Nnabuk O. Eddy, investigated that GC-MS study of ethanol extract of Andrographis paniculata revealed the presence of hydrocarbons ((E)-tetradec-3-ene), alcohol (6-methyl heptan-1-ol), aromatics (undecan-5-ylbenzene, dodecan5ylbenzene, undecan-2-ylbenzene, (2-methylundecan-6-yl) benzene and dioctyl phthalate). Some of the identified compounds were found to be green raw materials for polymer industries and pharmaceutical companies. Corrosion inhibition potentials of the ethanol extract of Andrographis paniculata was also investigated and it was found that the extract inhibited the corrosion of mild steel in HCl solution through the mechanism of physical adsorption. Quantum chemical approaches have been used to investigate the active components of the extract that are responsible for the inhibition of the corrosion of mild steel in HCl solution. The results obtained, indicated that the most active ingredient responsible for the corrosion inhibition process is pentadecanoic acid. Other compounds with lesser activity include 6-methyl hepta-1-ol, methyl palmitate, methyl icosanoate, methyl stearate, oleamide, dicotyl phthalate and ethyl docosanoate are the active.

40. Quantum chemical calculations using the Density Functional Theory (DFT) method at three different basis sets, namely, 6-31(d,p), 6-31+(d,p) and 6-311G(d,p) were performed on selected triazole and benzimidazole derivatives, namely 2-aminobenzimidazole (ABI), 1,3-benzothiazole (BTH), benzotriazole (BTA), 2-methylbenzimidazole (MBI), 2-(2-pyridyl)benzimidazole (PBI), 2-(amino methyl)benzimidazole (AMBI), 5-amino-3-mercapto-1,2,4-triazole (5AMTZ), 2- hydroxybenzimidazole (HBI), benzimidazole (BI) and 5-amino-1,2,4-triazole (5ATZ), to determine their reactive centres which might interact with the metal surface on the adsorption of the these compounds onto the metal surface. This was investigated by **Mwadham M Kabanda *et al.*,** The results show that the adsorption of the inhibitor onto the metal surface would preferentially be through the benzene ring that is fused to the heterocyclic ring and through the heteroatoms of the heterocyclic ring. The study on the protonated species of the studied compounds show that they have the least tendency to chemically adsorb onto the metal surface and might preferentially adsorb physically. The quantitative structure activity relationship approach indicates that three to four quantum chemical parameters are needed to effect a reasonable correlation between experimentally determined and theroetically estimated inhibition efficiencies.

41. The inhibition potentials of 2-(2-oxoindolin-3-ylideneamino) acetic acid (IG) and Indoline-2,3-dione(IS) was carried out using gravimetric and gasometric methods while the theoretical approach was investigated by **Nnabuk O. Eddy** ***et al.,*** using quantum chemical principles. The results obtained, indicated that IG and IS inhibited the corrosion of mild steel in HCl solution through the mechanism of chemical adsorption. The adsorption of the inhibitors is spontaneous, endothermic and is consistent with El Awardy et al kinetic model. Application of Dubinin-Radushkevich adsorption isotherm also ascertained the prevalence of chemisorption mechanism. The Fukui function, global softness and Huckel charges on carbon and electronegative elements in IG and IS revealed that the site for electrophilic attack is in the pyrrole nitrogen while the site for nucleophilic attack is in the phenyl ring. It was also found that the inhibition potential of IG is better than that of IS because the presence of methylenecarbamic acid group in IG (=NCH2CO2H) facilitated better adsorption than the =O group in IS.

42. P.Udhayakala ***et al*** .,were studied that the corrosion inhibition characteristics of two vinyl imidazole derivatives, ie., 2,4,5-triphenyl-1-vinyl-1*H*-imidazole (C1) and 2-(4-methoxyphenyl)-4,5-diphenyl-1-vinyl-1*H*-imidazole(C2), on mild steel by using Density functional theory (DFT).Quantum chemical parameters such as EHOMO (highest occupied molecular orbital energy), ELUMO (lowest unoccupied molecular orbital energy), the energy gap(ΔE), hardness(η), Softness(S), dipole moment(μ), electron affinity(EA), ionization potential(IE), the absolute electronegativity (χ) , electrophilicity index(ω) and the fraction of electron transferred (ΔN) have been calculated using B3LYP/6-31G(d,p) basis set. The local reactivity has been studied through the Fukui and condensed softness indices in order to predict both the reactive centres and to know the possible sites of nucleophilic and electrophilic attacks. The obtained correlations and theoretical conclusions agree well with the experimental data.

43. N. O. Eddy ***et al.,*** were investigate the inhibitory and adsorption potentials of 2-(4-methoxybenzylideneamino) acetic acid (MBDAC), 2-hydroxy-3-methoxybenzaldehyde (HMBA) and 4-methoxybenzaldehyde (MBA), were studied using experimental (gravimetric and gasometric) and computational chemistry simulation approaches. The results obtained from the experimental approach revealed that the trend for the inhibitory potentials of the three anisoles is HMBA > MBDAC > MBA. The observed trend is attributed to the structural differences resulting from the introduction of –OH, =O and NCH_2CO_2H substituents into the respective compounds. The adsorption of the three inhibitors on mild steel surface is spontaneous and supported the mechanism of chemical adsorption. Computational chemistry simulations was carried out by comparing calculated quantum chemical parameters (the energy of the frontier molecular orbitals, the energy gap, dipole moment and logP) of the studied anisoles in gas and aqueous phases with experimentally obtained % inhibition efficiencies. The tests gave excellent correlations for PM6, PM3, AM1, RM1 and MNDO Hamiltonians. Correlations between experimental and predicted inhibition efficiencies were also excellent. Local selectivity study using the Fukui function and global softness indices indicated that the likely sites for electrophilic and nucleophilic attacks are in the enol bonds of the inhibitors.

44. M.Abdallah *et al.,* were examine the inhibition effect of four synthesized novel bis(aminothiazole) derivatives on the corrosion of carbon steel in 0.5M H_2SO_4 solution was investigated using galvanostatic, potentiodynamic anodic polarization and weight loss techniques. The inhibitive effect was ascribed to the formation of insoluble complex adsorbed on the steel surface and the adsorption process follows Langmuir adsorption isotherm. A theoretical study of the corrosion inhibition efficiency of these bis(aminothiazole) derivatives, was carried out using density functional theory (DFT) at the B3LYP/6-31G(d) level of theory. Molecular properties related to the inhibition efficiency of these inhibitors were obtained and found in good correlation with the inhibition efficiency obtained from different techniques.

45. 1,5-Dimethyl-4-((2-methylbenzylidene)amino)-2-phenyl-1*H*-pyrazol-3(2*H*)-one (DMPO) was synthesized to be evaluated as a corrosion inhibitor by **Sutiana Junaedi *et al*.** The corrosion inhibitory effects of DMPO on mild steel in 1.0 M HCl were investigated using electrochemical impedance spectroscopy (EIS), potentiodynamic polarization, open circuit potential (OCP) and electrochemical frequency modulation (EFM). The results showed that DMPO inhibited mild steel corrosion in acid solution and indicated that the inhibition efficiency increased with increasing inhibitor concentration. Changes in the impedance parameters suggested an adsorption of DMPO onto the mild steel surface, leading to the formation of protective films. The novel synthesized corrosion inhibitor was characterized using UV-Vis, FT-IR and NMR spectral analyses. Electronic properties such as highest occupied molecular orbital energy, lowest unoccupied molecular orbital energy (E_{HOMO} and E_{LUMO}, respectively) and dipole moment (μ) were calculated and discussed. The results showed that the corrosion inhibition efficiency increased with an increase in the E_{HOMO} values but with a decrease in the E_{LUMO} value.

46. Kabanda *et al.*, tested on selected triazole and benzimidazole derivatives, by quantum chemical calculations using the Density Functional Theory method at three different basis sets, namely, 6-31(d,p), 6-31+(d,p) and 6-311G(d,p), to determine their reactive centres which might interact with the metal surface on the adsorption of the these39compounds onto the metal surface. The results show that the adsorption of the inhibitor on to the metal surface would preferentially be through the benzene ring that is fused to the heterocyclic ring and through the heteroatoms of the heterocyclic ring. The study on the protonated species of the studied compounds show that they have the least tendency to chemically adsorb onto the metal surface and might preferentially adsorb physically. The quantitative structure activity relationship approach indicates that three to four quantum chemical parameters are needed to effect a reasonable correlation between experimentally determined and theoretically estimated inhibition efficiencies.

47. Abdallah *et al.*, studied the inhibition effect of four synthesized novel bis (aminothiazole) derivatives on the corrosion of carbon steel in 0.5M H2SO4 solution. The inhibitive effect was ascribed to the formation of insoluble complex adsorbed on the steel surface and the adsorption process follows Langmuir adsorption isotherm. A theoretical study of the corrosion inhibition efficiency of these bis(aminothiazole) derivatives, was carried out using density functional theory (DFT) at the B3LYP/6- 31G(d) level of theory. Correlation coefficient matrix built to explore the dependence between the different molecular descriptors and the IE has good positive correlation coefficients with softness(S) and the fraction of electron transferred (ΔN) and good negative correlation coefficients with frontier orbital energy gap (ΔE) and the global hardness (η). The order of the inhibition efficiency of the inhibitors at different concentrations is given as follows: o-Bis-MeAT > o-Bis-AT > p-is-MeAT > p-Bis-AT. It is noted that the o-bisaminothiazole compounds are more efficientthan p-bisaminothiazole compounds attributed to the ortho effect.

48. The adsorption behaviour and inhibition mechanism of 2-aminomethyl benzimidazole(ABI), bis(2-benzimidazolylmethyl) amine (BBIA) and tri-(2-benzimidazolylmethyl) amine(TBIA) on the surface of mild steel were studied by **Cao *et al.,*** employing molecular dynamics (MD) simulations performed with Dmol3 module in Materials Studio software 6.0. It was found that the three molecules show the similar ability to donate electrons while the difference in inhibition performance should mainly be attributed to the difference in accepting electrons. MD simulations show that steric effect between the benzimidazole segments significantly affects the adsorptive configurations of the molecules on Fe (100) surface.

49. Obot *et al.,* studied the adsorption properties of three pyrazine compounds for steel corrosion was investigated by quantum chemical calculation and quenched force-field molecular dynamics simulation employing DFT calculations carried out using B3LYP 6-311G++ (d, p) basis set within the Gaussian-03 program. Global reactivity parameters such as 40EHOMO, ELUMO, the energy gap (ΔE) dipole moment showed close correlation with IE in the neutral molecules. The HOMO and LUMO orbitals are localized on the pyrazine rings which revealed the active sites responsible for the electrophilic and nucleophilic reaction of the inhibitor molecule with the steel. The Mulliken population analysis and the Fukui functions showed that the –C=N- centres were the main adsorption sites. The equilibrium adsorption behaviour of these pyrazine molecules on Fe (110) surface was investigated using molecular dynamics simulation. The binding energy followed the order: ABP > AP > MP, which is in agreement with experimental results.

50. Aouniti *et al.,* investigate the inhibitive performance of various amino acids experimentally. Methionine, cysteine and cystine who have a sulphur atom in their molecular structure are the best inhibitors. The corrosion inhibition of methionine is regarded by simple blocked fraction of the electrode surface related to the adsorption of inhibitor species according to Frumkin isotherm model on the Armco iron surface. Geometrical parameters for the investigated amino acids are optimised both in gas and aqueous phases, employing GGA Within the software package DMol3 in Materials Studio of Accelrys Inc using BLYP basis set. The computational method has proved satisfactory for the inhibition efficiency stimations.

51. Junaedi *et al.*, synthesized 1,5-Dimethyl-4-((2-methylbenzylidene)amino)-2-phenyl-1H-pyrazol-3(2H)-one and evaluated its efficiency as corrosion inhibitor on mild steel in 1MHCl using EIS, potentiodynamic polarization, OCP and EFM. The results showed that adsorption of DMPO onto the mild steel surface, leading to the formation of protective films. Electronic properties such as highest occupied molecular orbital energy, lowest unoccupied molecular orbital energy (EHOMO and ELUMO, respectively) and dipole moment (μ) were calculated using DFT within the program GAUSSIAN 03. The results showed that the IE increased with an increase in the EHOMO values but with a decrease in the ELUMO value and been reported that excellent corrosion inhibitors can not only offer electrons to unoccupied orbitals of the metal but also accept free electrons from the metal

52. Obi- Egbedi investigated on the adsorption and inhibition effect of xanthione (XION) on mild steel in 0.5M H2SO4 at 303–333 K by gravimetric and UV–visible spectrophotometric methods. Attempt to correlate the molecular structure to quantum chemical indices was made using density functional theory (DFT) at the B3LYP/6-31G (d) level using Spartan'06 V112 program package. Theoretical studies showed that the presence of sulphate ions with negative charges in the vicinity of the interface could favour the41adsorption of the positively charged inhibitor molecules through electrostatic interactions with the steel surface which now has negatively charged sulphate ions on it. Thus there is a synergism between SO4-2 ions and the protonated XION molecules

53. The corrosion inhibition of C38 steel in 1M HCl solution by 5-(2-chlorobenzyl)-2,6- dimethylpyridazin-3-one (CBDP) was studied by **Zarrok *et al.***, using weight loss measurement. The outcomes show that inhibition takes place by adsorption of the inhibitor on metal surface. Quantum chemical calculations using DFT at the B3LYP/6-31G* level of theory was further used to calculate some electronic properties of the molecule in order to ascertain any correlation between the inhibitive effect and molecular structure. Quantum chemical parameters such as EHOMO, ELUMO, ΔE, dipole moment (μ), number of transferred electrons (ΔN) and total energy (TE) were found to give good correlation with experimentally determined inhibition efficiency'

54. *John et al.,* reports the study on the inhibition of mild steel in 0.5 M HCl with substituted 1,2,4-triazoles (ACT & ATD) using conventional and quantum chemical calculations. Results show that the corrosion rate decreases and inhibition efficiency increases with inhibitor concentration. The theoretical study has been initiated using density functional theory at the B3LYP/6-31G (d) level in order to elucidate the inhibition efficiencies and reactive sites of these inhibitor molecules. The the local reactivity has been analysed through the Fukui function and condensed softness indices. The quantum chemical calculations and the electroanalytical results are in conformity with each other and the inhibitor molecule ACT acts as better inhibitor than ATD.

55. Bentiss *et al*., investigated on the relationships between electronic parameters and inhibition efficiency of oxadiazole, thiadiazole, and triazole derivatives. The results obtained for corrosion inhibitors of mild-steel demonstrated that correlation exists between inhibition efficiency and the energy band gap (ΔE). It was found that efficiency (related to0.001 M inhibitor concentration) is optimal at $\Delta E \cong 1.3$ beta. The efficiency depends, besides ΔE, on the conformation of the molecules, viz. planar (unsaturated) molecules are more efficient than partially saturated (and therefore twisted) derivatives

56. Liu *et al*., investigated on two 5-substituted tetrazoles, as copper corrosion inhibitorsin aerated 0.5 M H2SO4 solution by using potentiodynamic polarization, EIS and quantumchemical calculation employing DFT. The quantum chemical calculations show that energyof highest occupied molecular orbital (EHOMO), energy of lowest unoccupied molecular orbital42 (ELUMO), the energy band gap (ΔE) and dipole moment (μ) are found to correlate well withexperimental inhibition efficiencies of inhibitors

57. Gece *et al.,* investigated the corrosion inhibition efficiencies of 3-amino-1,2,4-triazole, 2-amino-1,3,4- thiadiazole, 5-(p-tolyl)-1,3,4- triazole, 3-amino-5-methyl mer capto -1,2,4-triazole and 2-aminobenzimidazole as corrosion inhibitors on steel in sodium chloride media using semi empirical PM3 and density functional theory (DFT) methods. Quantum chemical parameters have been calculated for these compounds by using semi empirical PM3and 6-31G(d), 6-311G(d, p) DFT methods. Through PM3 semi empirical and DFT quantum chemical calculations a correlation between parameters related to the electronic structure of some cyclic nitrogen compounds and their ability to inhibit the corrosion process could be established. The highest occupied molecular orbital energy levels and energy gaps calculated by 6-311G(d,p) DFT study show reasonably good correlation as compared to other calculated data. It was found that theoretical data support the experimental results.

58. Khaled ***et al.,*** suggested the use of cheminformatics tool for a prior dry-lab process, followed by a wet-lab process with contrast to the traditional techniques of identifying and synthesizing new corrosion inhibitors in wet lab. Quantum chemical method is used to explore the relationship between the inhibitor molecular properties and its inhibition efficiency. The density function theory (DFT) is also used to study the structural properties of three selected benzotriazole derivatives namely in aqueous phase. It is found that when the benzotriazole derivatives adsorb on the copper surface, molecular structure influences their interaction mechanism. The inhibition efficiencies of these compounds showed a certain relationship to highest occupied molecular orbital (HOMO) energy, Mulliken atomic charges and Fukui indices. A wet lab study has been carried out using weight loss, Tafel polarizationand impedance measurements to evaluate their inhibition performance in M HNO3. Results obtained from dry-lab process are in good agreement with those recorded from wet-lab experiments.

59. Khaled investigated on the inhibition performance of three triazole derivatives on mild steel in 1M HCl by various conventional methods. The adsorption behavior of these molecules at the Fe surface was studied by the molecular dynamics simulation method and the quantum chemical calculations. Results showed that these compounds inhibit the corrosion of mild steel in 1M HCl solution significantly. Molecular simulation studies were applied to optimize the adsorption structures of triazole derivatives. The iron/inhibitor/solvent interfaces were simulated and the charges on the inhibitor molecules as well as their structural43parameters were calculated in presence of solvent effects. Aminotriazole was the best inhibitor among the three triazole derivatives (triazole, aminotriazole and benzotriazole).

60. Larabi *et al.*, synthesied N-Phenyl oxalic dihydrazide and oxalic N-phenylhydrazideN,-phenylthiosemicarbazide and tested as inhibitors for the corrosion of mild steel in molar HCl by electrochemical measurements. The reactivity of these compounds was analysed through theoretical calculations based on density functional theory to explain the differentefficiency of these compounds as corrosion inhibitors. Low ELUMO and low energy gap (ΔE) explain the best inhibition efficiency of oxalic N-phenylhydrazide N,- Phenylthiosemicarbazide

61. A comparative theoretical study of three benzimidazole isomers, as corrosion inhibitors was performed by **Obayes *et al.*,** using density functional theory (DFT) with theB3LYP functional basis set. Nitro and amino groups were selected for investigation as substituents of the three corrosion inhibitors. Nitration of the corrosion inhibitor molecules led to a decrease in inhibition efficiency, while reduction of the nitro group to amine led to an increase in inhibition efficiency. These amino benzimidazole isomers represent a significant improvement in the inhibition efficiency of corrosion inhibitor molecules.

62. Eddy *et al.,* studied the experimental aspect of the corrosion inhibition potential of amino acid like adenine (AD), guanine (GU) and, hypoxanthine (HYP) using weight loss, potentiodynamic polarization and electrochemical impedance spectroscopy (EIS) methods while the theoretical aspect of the work was carried out by calculations of semi-empirical parameters (for AM1,MNDO,CNDO, PM3 and RM1 Hamiltonians), Fukui functions and inhibitor–metal interaction energies. Results obtained from the experimental studies were in good agreement and indicated that the amino acids are good adsorption inhibitors for the corrosion of aluminum in HCl. Quantum chemical studies revealed that the experimental inhibition efficiencies of the studied purines are functions of some quantum chemical parameters including total energy of the molecules (TE), energy gap (EL–H), electronic energy of the molecule (EE), dipole moment and core–core repulsion energy (CCR). Fukuifunctions analysis through DFT and MP2 theories indicated slight complications and physical results. However, results obtained from calculated Huckel charges, molecular orbital and interaction energies, the adsorption of the inhibitors proceeded through the imine nitrogen (N5) in GU, emanine nitrogen (N7) in AD and the pyridine nitrogen (N5) in HPY

63. The density functional theory (DFT) at the B3LYP/6-31G (d,p) basis set level method were performed on three hector bases used as corrosion inhibitors; namely, 3-anilino-5-imino-4-phenyl- 1,2,4-thiadiazoline (AIPT), 3-anilino-5-imino-4-tolyl-1, 2,4-

thiadiazoline(AITT), and 4-(4-chlorophenyl)-5-imino-N-phenyl-4,5-dihydro-1,2,4-thiadiazol - 3-amine(AICT) by **Khodaei Tehrani and Niazi.** They were used as corrosion inhibitors for mild steel in acidic medium in order to determine the relationship between molecular structure and their corresponding inhibition efficiency (%IE). The results of the quantum chemical calculations and experimental %IE were subjected to correlation analysis. This indicates that their inhibition effects are closely related to the highest occupied molecular orbital energy (EHOMO), the lowest unoccupied molecular orbital energy (ELUMO), the energy gap (ΔE), the hardness (η), the softness (σ), the electronegativity (ζ), and the fraction of electrons transferred from the inhibitor molecule to the metal surface (ΔN). In addition, the local reactivity has been analyzed through the Fukui function. Two QSAR equations were developed and used to predict the corrosion inhibition efficiency for hector bases

2.4 Drugs

64. Nnabuk O. Eddy and Eno E. Ebenso were investigated about the corrosion inhibition of mild steel in H_2SO_4 by Cloxacillin (CLX) and its synergistic combination with halides (KI, KBr and KCl) by using gravimetric, thermometric, gasometric and quantum chemical methods. The results indicated that CLX is a good adsorption inhibitor for the corrosion of mild steel in H_2SO_4. Inhibition efficiency of CLX on mild steel increased with increase in concentration but decreased with rise in temperature ranging from 90.40 to 93.98%, 80.85 to 86.43%, 78.41 to 84.07% and 58.61 to 77.57% at 303, 313, 323 and 333 K respectively. The combination of CLX with the halides (KCl, KBr and KI) enhanced the inhibitive and adsorption capability of CLX to a greater extent indicating synergism. The adsorption was spontaneous and the experimental data were also found to fit into the Langmuir (mean R2 = 0.9978) adsorption isotherm. Physical adsorption mechanism has been proposed from the thermodynamic data obtained in the study and the activation energies of the corrosion inhibition by CLX were higher than the value obtained for the blank which signifies physical adsorption. Some quantum chemical parameters and the Mulliken charge densities on the optimized structure of CLX were calculated using the B3LYP/6-31G (d,p) basis set method to provide further insight into the mechanism of the corrosion inhibition process. The local reactivity was analyzed through the Fukui function and condensed softness indices in order to know the possible sites of nucleophillic and electrophillic attacks.

65. Mwadham M. Kabanda ***et al.,*** were investicated the corrosion inhibitive effects of two naphthylazo anionic dyes namely Calmagite (CG) and Fast sulphone black F (FSBF) on mild steel surface in hydrochloric acid solution using electrochemical impedance spectroscopy (EIS),Tafel polarization techniques and quantum chemical calculation methods. Inhibition efficiency increased with increase in concentration of the inhibitors. Trends in the calculated molecular properties (e.g., dipole moment, HOMO and LUMO energies) were compared with trends in the experimentally determined inhibition efficiency. The results show that trends in the quantum chemical descriptors are in agreement with the experimentally determined inhibition efficiencies. The adsorption on the metal surface might be due to both chemisorption and physisorption mechanisms arising from the anionic nature of the molecules in solution (which promote physisorption) and the electron donating effects of the electron rich centres (which promotes chemisorption). The solvent effects have minimal influences on the characteristics of the calculated molecular properties of the dyes.

66. Clotrimazole (CTM) [1-[(2-chlorophenyl)-diphenyl-methyl] imidazole], an antifungal drug, was investigated as a corrosion inhibitor for aluminium in HCl using weight loss method by **I.B. Obot** ***et al.,*** CTM inhibited the corrosion of aluminium in HCl. The inhibition efficiency increased with increase in the concentrations of CTM to reached 90.90% at 1 x 10-4M, but decreased with increase in temperature. Phenomenon of physical adsorption is proposed for the inhibition and the process followed the Langmuir adsorption isotherm and kinetic / thermodynamic model of El-Awady et al. The mechanism of adsorption inhibition and type of adsorption isotherm were proposed from the trend of inhibition efficiency with temperature, Ea, _Gads, and Qads. Quantum chemical calculations results show that CTM possesses a number of active centres concentrated mainly on the imidazole moiety of the molecule. The highest occupied molecular orbital (HOMO), and the lowest unoccupied molecular orbital (LUMO) were also found around the nitrogen atoms and the cyclic of the benzene rings.

67. The effect of Metformin as Corrosion Inhibitor for Mild Steel in Hydrochloric Acid Solution was investigated by **Ambrish Singh** ***et al.,*** using gravimetric, potentiodynamic polarization and electrochemical impedance spectroscopy. Polarization results revealed that Metformin acted as mixed type inhibitors. The inhibition efficiency increases with increasing the inhibitors concentration and decreases with increasing the temperature. The gravimetric and potentiodynamic polarization results were in good agreement with the electrochemical impedance spectroscopy data.

68. The expired desloratidine drug was examined as an inhibitor for carbon steel (X52) corrosion in 1M HCl solution. The inhibition conduct of desloratidine was studied utilizing potentiodynamic polarization and weight loss techniques. The drug efficacy as an inhibitor was found to rise by increasing the desloratidine concentration and reduces by increasing temperature. Morphology of the surface and the adsorption isotherm were also studied. The heat of adsorption (Qads) and activation energy (Ea) were computed and elucidated. Based on adsorption of desloratidine molecules on carbon steel surface leading to conserve the carbon steel surface from the offensive ions, the inhibition conduct was explicated.

69. A. S. Fouda *et al.,* studied the effect of an antibacterial Septazole on the corrosion of copper in 0.1 M HCl at different concentrations using potentiodynamic polarization, electrochemical frequency modulation (EFM) and electrochemical impedance spectroscopy (EIS) measurements. Potentiodynamic polarization indicated that Septazole acted as mixed- type inhibitor. EIS measurements were also used to investigate the mechanism of corrosion inhibition. Results obtained from EFM technique were in agreement with potentiodynamic and EIS techniques. The adsorption of Septazole on the surface of copper followed the Langmuir adsorption isotherm at all concentrations and temperatures studied. The reactivity of Septazole was analyzed through quantum chemical calculations based on PM3 semi-empirical method to explain the efficiency of this inhibitor as corrosion inhibitors.

70. Inhibition performance of Lornoxicam & Tenoxicam against corrosion of carbon steel in 1M H_2SO_4 solutions was investigated by **A. S. Fouda *et al.,*** weight loss, potentiodynamic polarization and electrochemical impedance spectroscopy (EIS) measurements. The inhibition efficiency increased with increasing inhibitor's concentration, but decreased with increase in temperature. Potentiodynamic polarization curves showed that, the inhibitors were of mixed type. The apparent activation energy and other thermodynamic parameters for the corrosion process have also been calculated and discussed. The inhibition of carbon steel corrosion is due to the adsorption of the inhibitor molecules on the surface, which follows Temkin adsorption isotherm. The mechanism of inhibition was discussed in the light of the chemical structure of the undertaken inhibitors.

71. L. Magaji ***et al.,*** studied the corrosion inhibition potentials of ciprofloxacin were carried out using gravimetric, thermometric and scanning electron microscopy techniques. The inhibition efficiencies of ciprofloxacin for the corrosion of mild steel increased with increase in concentration but decreased with rise in temperature. The scanning electron microscopy shows that the damaging effect of the acid on the metal surface is greatly reduced with the presence of the inhibitor. This may probably be due to the formation of protective coverage by the inhibitor. Adsorption characteristics of the inhibitor has also been studied and it is found that ciprofloxacin inhibits the corrosion of mild steel by being adsorbed on the surface of mild steel by a physical adsorption mechanism. The adsorption of ciprofloxacin on the mild steel surface was found to be spontaneous and obey the Langmuir adsorption isotherm model.

72. .R. S. Abdel Hameed ***et al.,*** were investigated that Expired declophen(2-(2, 6-dichloranilino) phenylacetic acid) ampoules were evaluated as non- toxic corrosion inhibitors for mild steel in 1 M HCl solutions, by chemical and electrochemical techniques, at different concentrations and temperature. The corrosion inhibition efficiency increased with increase in inhibitor concentration, but decreased with increase in temperature. Potentiodynamic polarization studies indicate that expired Declophen drugs acts as a mixed type inhibitor. The outcomes show that inhibition takes place by adsorption of the inhibitor on metal surface without altering the mechanism of corrosion process. The adsorption of Declophen takes place according to Langmuir`s adsorption isotherm. Thermodynamic parameters were calculated and discoursed. The data collected from the studied techniques are in good agreement to confirm the ability of using expired Declophen ampoules as corrosion inhibitors for steel in hydrochloric acid corrosive medium.

73. The corrosion inhibition of mild steel in 1M HCl solution by Cefixime was studied ate temperature range 303–333K by Imran **Naqvi** ***et al.,*** using weight loss measurement and electrochemical techniques i.e. Polarization Resistance, Potentiodynamic Polarization and Electrochemical Impedance Spectroscopy (EIS). The results obtained at 30oC divulged that this pharmaceutical compound had established > 90% inhibition efficiency at an optimum concentration of 8.8×10−4M. The outcomes show that inhibition takes place by adsorption of the inhibitor on metal surface without altering the mechanism of corrosion process. The adsorption of Cefixime takes place according to Langmuir`s adsorption isotherm. Kinetic parameters (activation energy, and pre-exponential factor, λ) as well as thermodynamic parameters (enthalpy, entropy and free energy of adsorption; and respectively) were calculated and discoursed. Potentiodynamic polarization studies indicate that Cefixime acts as a mixed type of inhibitor. Data collected from EIS studies has been analyzed to model the appropriate equivalent circuit for better explanation of corrosion inhibition process. The inhibition mechanism of Cefixime has also been discussed.

74. Expired ranitidine was tested as a corrosion inhibitor by **R.S. Abdel Hameed** for mild steel in 1 M HCl using different techniques: weight loss, potentiodynamic polarization, open circuit potential and electrochemical impedance spectroscopy. The polarization resistance (Rp) value increased with increase in the concentration of the inhibitor. Results obtained revealed that ranitidine performed excellently as a corrosion inhibitor for mild steel in this medium at 303 K. The protection efficiency increased with increasing inhibitor concentration. The maximum protection efficiency of 90% has been obtained at 400 ppm. On the other hand, the efficiency decreases with increasing temperature. The adsorption of the inhibitor on the mild steel surface followed Langmuir adsorption isotherm model. The activation and thermodynamic parameters of dissolution and adsorption were calculated and discussed. The negative value of _Gads (-40 kJ mol-1) indicates spontaneous chemical adsorption. Results obtained from polarization, EIS and weight loss measurements are in good agreement with each other.

75. The corrosion inhibition action of sulphadoxine and pyrimethamine, which is being sold under the trade name "Antimal" as a single antimalarial drug was used as low cost and eco-friendly inhibitor for acid corrosion of mild steel monitored by **I. A. Akpan and N. O. Offiong** using an electrochemical technique at room temperature (30°C). It was found that the polarization resistance of the inhibitor was directly proportional to the concentration of the inhibitor –making the metal more resistant to corrosion at higher inhibitor concentrations. Also, the inhibition efficiency was found to increase with rise in concentration of the inhibitor. The experimental data were found to fit better for Temkin isotherm which illustrates multilayer chemisorption. However, significant correlation (R2 = 0.76) was found for Frumkin isotherm indicating possible amphiphilic interactions from alkyl substituents on the inhibitor molecules.

76. This research demonstrates the effect of some amide compounds (1 and 2) as corrosion inhibitors on C-steel in 1.0 M of hydrochloric solutions utilizing mass reduction studies, electrochemical [potentiodynamic (PP), AC impedance measurements (EIS), electrochemical frequency modulation] techniques, and surface checks were used to illustrate the importance of amide compounds to the corrosion protection process of C-steel. The tests displayed that the inhibition efficiency (IE %) augmented with increase in amide dose but reduced with the growth of temperature. The highest inhibition efficiency is 99% for compound (1) and 98.8% for compound (2) at the maximum dose tested (50 μ the M) by polarization method. PP data show that these compounds affect both cathodic and anodic processes (i.e. mixed type) and were adsorption on the carbon steel obeying Langmuir adsorption isotherm. The EIS results indicate that the changes in impedance parameters are related to the adsorption of amides on the alloy surface. Scanning of electron microscopy, energy transmitted X-ray—Atomic force microscopy studied the morphology of inhibited C-steel.

77. The influence of three compounds of anti-ulcer prazole drugs on the dissolution of aluminum electrode in 1.0M HCl solution was investigated by **Omar A. Hazazi and M.Abdallah** using weight loss, hydrogen evolution reaction, galvanostatic polarization and electrochemical impedance spectroscopy techniques .The inhibition efficiency was found to increase with increasing the concentration of the inhibitors and with decreasing the temperature. The inhibition action of these compounds was explained in terms of adsorption on the aluminum surface. The adsorption process follows Langmuir isotherm. Some activated thermodynamic parameters were calculated. The polarization studies showed that these compounds are acting as mixed inhibitors. Electrochemical impedance spectroscopy technique exhibit one capacitive loop indicating that, the corrosion reaction is controlled by charge transfer process. The results obtained from different techniques are in good agreement.

78. Nicolae Vaszilcsina ***et al.,*** was used a method of expired or unused drugs valorization as corrosion inhibitors for metals in various media. Cyclic voltammograms were drawn on platinum in order to assess the stability of pharmaceutically active substances from drugs at the metal–corrosive environment interface. Tafel slope method was used to determine corrosion rates of steel in the absence and presence of inhibitors. Expired Carbamazepine and Paracetamol tablets were used to obtain corrosion inhibitors. For the former, the corrosion inhibition of carbon steel in 0.1 mol L−1 sulfuric acid solution was about 90%, whereas for the latter, the corrosion inhibition efficiency of the same material in the 0.25 mol L−1 acetic acid–0.25 mol L−1 sodium acetate buffer solution was about 85%.

79. Due to the toxicity of widely used corrosion inhibitors and the ever tightening environmental regulations surrounding their use and disposal, there is great interest in replacing harmful inhibitors with effective non-hazardous alternatives. Over the past two decades, extensive research and development have led by **Gokhan Gece** ***et al.***, to the discovery of new classes of corrosion inhibitors, and the importance on the use of several drugs as corrosion inhibitors has grown. This review presents most of the contributions made to the literature on the use of drugs as corrosion inhibitors of various metals in recent years.

80. The adsorption and corrosion inhibition behaviour of ceftezole (CFT) at mild steel surface were studied by **Ashish kumar sing and Eno E. Ebenso** using gravimetrically and electrochemical impedance spectroscopy (EIS) and tafle polarization technique. The increase in concentration, immersion time and temperature shows a positive effect on inhibition efficiency. Inhibition molecules directly adsorbed on the surface on the basis of donor acceptor interacts between the p-electrons of iron atoms. The adsorption of CFZ followed Langmuir adsorption isotherm. Potentiodynamic polarization study revealed that CFZ acted as mixed type of inhibitor. The results obtained from different methods are in good agreement.

81. Ahmad M. El-Desoky ***et al.,.*** were state that the expired pharmaceutical compounds such as Hydrochlorothiazide (HCT), captopril (CAP) and guaifenesin (GFN) were can used as corrosion inhibitors. These drugs were evaluated as corrosion inhibitors for C- steel in 2 M HCl solution at 30°C by various analytical techniques. Electrochemical impedance spectroscopy (EIS), potentiodynamic polarization, electrical frequency modulation (EFM) and weight loss technique data were discussed. The electrochemical results indicated that these compounds are efficient inhibitors for C- steel corrosion. Their efficiencies were up to 81.8 % at the tested temperature. The increase in drug concentration caused increase in the corresponding inhibition efficiency (%IE) values. The adsorption amount and the coverage of pharmaceutical molecules were also increased with increasing concentration. The adsorption of these compounds on C- steel surface obeyed the Temkin adsorption isotherm. Polarization curves indicated that, the investigated pharmaceutical compounds were mixed-type inhibitors. The corrosion kinetic parameters were discussed. The chemical data were in good agreement with those obtained from electrochemical methods.

82. The effectiveness of Expired Voltaren (EV) drug as corrosion inhibitor for aluminium in 1 M HCl was investigated by **R. S. Abdel Hameed *et al.*,** using weight loss and electrochemical methods. Polarization curves revealed that the inhibiting action of the EV is mixed-type. The addition of the inhibitor molecules shifted the corrosion potential (E_{corr}) to more noble direction. The influence of temperature on the adsorption of inhibitor onto aluminium surface has been reported. Activated thermodynamic parameters were computed and discussed. The adsorption of the EV was found to obey the Langmuir adsorption isotherm. The inhibition efficiency of expired Voltaren, EV obtained by all methods was in good agreement with each other.

83. Narayana Hebbar *et al.,* were investigated the Corrosion inhibition efficiency of Ketosulphide analysed by AC-impedance method and Polarisation method. The parameters such as corrosion potential (E_{corr}), corrosion current density (i_{corr}), cathodic Tafel slope (βc), anodic Tafel slope (βa) is computed by Tafel method. The R_p value and Cdl values are obtained by AC-impedance method. There is a good agreement between the AC- impedance method and polarization method. The percentage inhibition efficiency of ketosulfide is 75% in IM HCl solution at 303K. The decrease in the current density and the corrosion rate is decreased with the concentration of the ketosulfide and also the polarization method showed the mixed type inhibition characteristics of Ketosulphide. The increase in the Rp value with the increase in the concentration of inhibitor is calculated by the AC-Impedence method showed that the Ketosulphide is good inhibitor. There is a decrease in the Cdl value with the concentration which supports the electrochemical results.

2.5 Paint

84. Electrochemical impedance spectroscopy(EIS) in the 100 KHz-10 mHZ frequency range was employed as the main electrochemical technique to study the corrosion protection behaviour of zinc rich epoxy paint in 3% NaCl solution. **Nadia Hammouda *et al*** ., were studied that the EIS results obtained at the open-circuit corrosion potential have been interpreted using a model involving the impedance impedance of particle to particle contact to account for the increasing resistance between zinc particles with immersion period, in addition to the impedance due to the zinc surface oxide layer and electrical resistivity of the binder. Galvanic current and dc potential measurements allowed us to conclude that the cathodic protection effect of the paint takes some time to be achieved. The loss of cathodic protection is due to a double effect: the decrease of the Zn/Fe area ratio due to Zn corrosion and the loss of electric contact between Zn to Zn particles. Even when the cathodic protection effect by Zn dust become weak, the substrate steel is still protected against corrosion due to the barrier nature of the ZRP film reinforced by Zn.

85. B.M. Durodola *et al.*, was investigated the corrosion resistance of electrolytic Zinc-coating on mild steel in sea water. Mild steel samples were zinc electroplated at voltage varied (0.6, 0.8, 1.0, 1.5V) and for plating time also varied (20, 30,40,50,60 minutes).Focused ion-beam scanning electron microscopy was used to observe the surface morphology of the plated surface. The visual observation of plated samples revealed that samples plated at 0.6V for 20 and 60 minutes: 0.8V for 30 minutes and 1.0V for 40 minutes showed the best physical properties such as Colour, brightness, fineness, streak and strong adhesion. The plated samples were then subjected to sea water environment for 30 days. The electro-potentials mV (SCE) was measured daily. Weight loss was determined at intervals of five days for the duration of the exposure period. Experimental result showed little corrosion attack on the zinc-plated sample on the fifteenth and twenty-fifth days exposure. The severity of the attack decrease with increasing weight of zinc coated on the substrate. Significant increase in the corrosion resistance were achieved for the plated samples compared with that of the substrate. The pH of the sea water became more alkaline at the end of the thirty days test.

86. A conductive paint coating was prepared by **Ahmad Azmin Mohammed *et al.,*** using a mixture of cathode waste materials (CWM) as the filler and epoxy resin as the binder. The optimum electrical conductivity of 5.33*10-4 Scm-1 was obtained at 25wt% CWM. It was found that the contact of the filler improves the electrical conductivity. However, the corrosion Potential (Ecorr) values of paint/mild steel decreased with the higher percentage of CWM to the paint. The low percentage incorporation of the filler shows the best ability to protect the mild steel from corrosion among coatings.

87. K.L.Palanisamy *et al.,* was investigated that the corrosion inhibitors are chemical substances that are added to the corrosive environment to reduce or eliminate corrosion. The good inhibitor has many advantages such as high inhibition efficiency, low price, low toxicity and easy production. Nano material exhibit such advantages with addition of improvement of the environmental impact. The stabilization is the significant factor for these iron oxide nano particlest exhibit high corrosion inhibition efficiency. Hence it was decided to have a better corrosion protection for these metals, iron oxide nano particles (IONPs) functionalized with carrier oils such as olive oil and lineed oil as natural stabilizing agents composited with paint after six months of storage at room temperature and coated on mild steels. The morphology of these six months stored IONPs was investigated by scanning electronic microscopy (SEM) the weight loss method was chosen for determining the corrosion inhibition efficiency and corrosion rate.

88. This study involves the productive effect of organic coating used on reinforced concrete in an acid environment similar to this existing in certain metallurgic processes in which the concrete degradation and the steel corrosion cause significant cost to the metal production which is studied by **Rosa Vera *et al.,*** A 0.50 w/c concrete was prepared and then characterised through physical and mechanical test such as electrical resistance, capillary absorption, total absorption, total porosity, chloride permeability and densile and compressive strength. The uncoated acrylic and epoxy coated concrete cylinders where exposed to an artificial acidic solution for 589 days the corrosion of the steel reinforcement was evaluated by means of correction potential and polarization resistance. The morphology of the corrosive attack was observed using scanning electron microscopy (SEM) and the composition of the corrosion products was determined using an X ray diffractometer (XRD) the deterioration of the concrete and its level of contamination where evaluated by measuring the concentration pro=files of chloride acidity and sulphate. The results show that the reinforcing steel reached an active state in the uncoated concrete and in the acrylic coated concrete, where as it remained in a passive state in the epoxy coated concrete. Under the conditions established in this study and based on chemical analysis and on the electrochemical test, it was determined that the coating which performed better as a barrier in an acidic medium was the epoxy coating.

89. A. Cook, A. Gabriel and N. Laycock were investigated the protective capacity of solution caused polyaniline coating systems toward mild steel has been investigated via simple electrochemical polarization techniques. Coatings containing a polyaniline-emeraldine sault species were found to significantly reduce corrosion rate in 0.1 mol dm^{-3}NaCl and 0.1 mol dm^{-3} HCL. The non-conduction counterpart, polyaniline-emeraldine base also provided some degree of production in the latter medium. Most importantly, the data confirms that an anodic protection scheme cannot operate in such media. The results presented herein support the emerging paradigm that protection stems from the inhibitory properties of the polymer's dopant species.

2.6 Coating

90. Adriana Samide *et al.,* investigated the study aims to implement a strategy to reduce the carbon steel corrosion rate in a sulfuric acid solution, using an expired drug with adsorption affinity on the metal surface. To investigate the corrosion protection efficiency of an environmentally friendly inhibitor, namely neomycin sulfate (NMS), the electrochemical measurements were applied on carbon steel immersed in 1.0 M H2SO4 solution with and without NMS. The protective layer that formed on the steel surface was studied by atomic force microscopy (AFM). The potentiodynamic polarization and electrochemical impedance spectroscopy (EIS) showed that the presence of the neomycin sulfate in acid solution leads to the decrease in corrosion current density (icorr) and the increase of polarization resistance (Rp). The mixed mechanism between physical and chemical adsorption of NMS molecules on the steel surface was proposed according to the Langmuir adsorption isotherm. AFM indicated that the NMS molecules contributed to a protective layer formation by their adsorption on the steel surface. The AFM parameters, such as root mean square roughness (Rq), average roughness (Ra), and maximum peak to valley height

(Rp−v) revealed that in the presence of NMS a smoother surface of carbon steel was obtained, compared to the steel surface corroded in sulfuric acid blank solution.

91. Chandrabhan Verma ***et al.,*** investigated that due to the increasing ecological awareness and environmental regulations,recent advancements in the field of corrosion science and technology isdirected towards "green chemistry". One of the major factors that restrict the utilization of synthetic corrosion inhibitors is their toxic nature. The preparation of organic corrosion inhibitors is associated with the huge discharge of environmentally malignant chemicals into the surrounding environment. Commercially drugs present ideal candidature to replace these traditional toxic corrosion inhibitors. The article overviews the results of previous works describing the influence of chemical medicines on the corrosion inhibition of industrially important metals in acid solution.

92. Menaga ***et al.***, investigated that the inhibitive ability against mild steel corrosion of pharmacologically active compound Bexol towards acid was valued by gravimetric experiments and electrochemical experiments. The results of the weight loss studies indicated inhibition efficiency was boosted with increased concentration of Bexol from 5 mg/L to 45 mg/L in the acid environment. Potentiodynamic polarization behaviour of the metal specimen in presence of Bexol exposed the mixed style inhibitive action. EIS results established that Bexol adsorb at the metal/acid junction by mere electrostatic physical adsorption mechanism to inhibit the corrosion process. Epoxy coated mild steel without and with the studied bexol was used for electrochemical studies in 3.5% of NaCl for different time intervals. The Density Functional theory (DFT) was employed to scrutinize inhibition property of the considered inhibitor. Experimental measurement obtained shows perfect fit to the theoretical calculations.

93. Lekan Taofeek Popoola ***et al.,*** reported that lives and wealth have been lost due to corrosion in almost all engineering fields. Not only this, the cost of reviving damaged equipment's in the industry due to corrosion contributed a lot to the gross domestic product of a nation. Thus, all hands must be on desk to combat this hazardous act via time-to-time research on its final resolution. However, current research works have revealed effective and reliable corrosion inhibitors from pharmaceutical drugs, plant extracts and ionic liquids as organic green corrosion inhibitors (OGCIs) with accommodative attributes such as being environmentally friendly, readily available, biodegradable, non-harmful, relatively cheap and many others to mention a few. This paper opens readers mind into the detailed classifications, mechanisms and active functional groups of these eco-friendly OGCIs. Not only the corrosion efficiency calculation ways but also influencing factors on efficiency were presented. Plant extracts, pharmaceutical drugs, ionic liquids and synthetic inhibitors, as among major sources of OGCIs, used in preventing material corrosion in corrosive media were separately and comprehensively examined. The significance of values obtained from simulating presented mathematical models governing OGCIs kinetics, adsorption isotherm and adsorption thermodynamics was also included. In conclusion, beneficial recommendations for both current and prospective researchers in the field of Corrosion Engineering were presented.

94. Xihua Xu *et al.,* The corrosion mitigation efficiency of biotin drug for mild steel in 15% hydrochloric acid was thoroughly investigated by weight loss and electrochemical methods. The surface morphology was studied by the contact angle, scanning electrochemical microscopy, atomic force microscopy and scanning electron microscopy methods. Quantum chemical calculation and Fukui analysis were done to correlate the experimental and theoretical data. The influence of the concentration of inhibitor, immersion time, temperature, activation energy, enthalpy and entropy has been reported. The mitigation efficiency of biotin obtained by all methods was in good correlation with each other. Polarization studies revealed that biotin acted as a mixed inhibitor. The adsorption of biotin was found to obey the Langmuir adsorption isotherm. Surface studies showed the hydrophobic nature of the steel with inhibitor and vindicated the formation of a film on the metal surface that reduced the corrosion rate.

95. Umoren *et al.,* reported that the use of corrosion inhibitors is the one the most economical and practical means of controlling metallic corrosion in different corrosive media. Interest on the use of eco-friendly organic compounds as corrosion inhibitors has extended to the use of polymers to inhibit metallic corrosion. The use of polymers as corrosion inhibitors has attracted attention and has gained wide acceptance in recent times. This has been attributed to a number of factors namely: (i) they are low cost and stable to metallic materials in acid media, (ii) they possess multiple adsorption sites, (iii) through their functional groups, they form complexes with metal ions and on the metal surface these complexes occupy a large area, thereby blanketing the surface and protecting the metal from corrosive agents present in the solution. A number of synthetic and natural polymers as metallic corrosion inhibitors have been investigated and reported.

96. Pang xuehui ***et al.*****,** investigated that corrosion inhibition of mild steel in 1 mol/L HCl by amoxicillin and ceftriaxone in the concentration range of 1.00×10−5−1.00×10−2 mol/L has been studied using weight loss, electrochemical potentiodynamic polarization, electrochemical impedance spectroscopy (EIS), scanning electron microscopy (SEM) and quantum chemistry tests at 298 K. The weight loss experiment showed that the inhibition efficiency increased with amoxicillin and ceftriaxone concentrations to attain the maximums of 80.3% and 94.1%, respectively at 1.00×10−2 mol/L. Potentiodynamic polarization indicated that amoxicillin and ceftriaxone acted as mixed-type inhibitors but mainly inhibited cathode hydrogen evolution reaction for mild steel in 1 mol/L HCl. The electrochemical impedance spectroscopy (EIS) demonstrated the inhibitors covered the active points of metal surface to inhibit corrosion. The absorption of both inhibitors on the mild steel surface was found to follow Langmuir adsorption isothermal and dominantly involve chemical adsorption at 298 K. Scanning electron microscopy (SEM) confirmed both of the inhibitors played a significant protective effect in mild steel corrosion in 1 mol/L HCl. The relationship between the inhibition properties and molecular structure had been discussed by quantum chemistry calculation. All the experimental results concluded that both amoxicillin and ceftriaxone acted as good corrosion inhibitors and their inhibition efficiency was in the order of ceftriaxone > amoxicillin.

97. Mene Benmahammed ***et al.*****,** investigated that the study of the corrosion inhibition of mild steel in acid medium 1M HCl by the Schiff base compounds named {4,4'-Bis(pyrrole-2-carboxaldehyde) diphenyl diimino sulfide (L1) and 4,4'-Bis(thiophene-2-carboxaldehyde) diphenyl diimino sulfide (L2)} was carried out using various techniques: weight loss measurements, polarization curves, electrochemical impedance spectroscopy (EIS), atomic force microscopy (AFM),Scanning electron microscopy (SEM) and energy dispersive X-ray spectroscopy (EDS).The influence of the concentration, immersion time and temperature were examined and the mode of adsorption of these inhibitors on the surface of the metal was highlighted by assigning the appropriate isotherm. The experimental results indicate that these compounds are effective corrosion inhibitors and the inhibitory competence rises with increasing inhibitor concentration. The adsorption of these compounds on the mild steel surface obeys the isotherm of Langmuir. The correlation between the molecular structures and the inhibitory properties of the compounds studied was performed using the Density Functional Theory method (DFT). Furthermore, molecular dynamic simulation (MD) has been taken into account. The results indicate that the adsorption energy of L1 was less than L2, which is in accordance with the experimentally determined inhibition effect.

98.Akhil Saxena *et al.*, found that the corrosion hindrance impact of Syzygium aromaticum fruits extract has been analyzed on mild steel corrosion in 0.5 M H2SO4 by utilizing weight reduction estimations, potentiodynamic polarization estimations and electrochemical impedance spectroscopy (EIS) procedures. This Eugenol and Eugenol acetate containing extract diminishes the corrosion rate of mild steel in acidic medium. The greatest corrosion restraint effectiveness was observed at 500 mg/L inhibitor concentration in 0.5 M H_2SO_4. The adsorption of Syzygium aromaticum extract on the surface of mild steel has been analyzed by utilizing AFM, SEM study and spectroscopic strategies.

99.Parul Dohare *et al.*, investigated that the effectiveness of expired Tramadol (ETr) as a corrosion inhibitor for mild steel in 1M HCl solution was tested by electrochemical impedance spectroscopy (EIS), potentiodynamic polarization (PDP), surface analysis (SEM and AFM), and density functional theory (DFT) methods. The data obtained from EIS and PDP curve indicated the inhibition efficiency is in direct relationship to Tramadol concentration. The Langmuir adsorption isotherm data confirms a physiochemical mechanism. The SEM and AFM analyses reveal a smoother surface of mild steel. The DFT calculations were used to evaluate the electronic parameters of the neutral and protonated forms of the drug.

References

[1] Y. Meng, W. Ning, B. Xu, W. Yang, K. Zhang,a Y. Chen, L. Li, X. Liu, J. Zhenga and Y. Zhangb, "Inhibition of mild steel corrosion in hydrochloric acid using two novel pyridine Schiff base derivatives: a comparative study of experimental and theoretical results," *Royal Society of Chemistry Advances*, vol. 7, pp.43014–43029, 2017

[2] I. O. Arukalam, I. O. Madu, N. T. Ijomah, C. M. Ewulonu, and G. N. Onyeagoro, "Acid Corrosion Inhibition and Adsorption Behaviour of Ethyl Hydroxyethyl Cellulose on Mild Steel Corrosion," *Hindawi Publishing Corporation Journal of Materials,* vol.2014.

[3] Dakeshwar Kumar Verma, "Corrosion Inhibition of Mild Steel by Extract of Bryophyllum Pinnatum Leaves in Acidic Solution," *Chemistry and Materials Research,* vol.7, no.5, 2015.

[4] N. Palaniappana, J. Alphonsab, I.S. Colec, K. Balasubramani and, I.G. Bosco, "Rapid investigation expiry drug green corrosion inhibitor on mild steel in NaCl medium," *Materials Science & Engineering* B 249, pp.114423, 2019

[5] A. Khamis, M.M. Saleh, M.I. Awad," Synergistic inhibitor effect of cetylpyridinium chloride and other halides on the corrosion of mild steel in 0.5 M H2SO4," *Corrosion Science*, vol.66, pp.343-349,2013

[6] Jamiu K. Odusote, and Olorunfemi M. Ajayi, "Corrosion Inhibition of Mild Steel in Acidic Medium by Jathropha Curcas Leaves Extract," *Journal of Electrochemical Science and Technology*, vol. 4, no. 2, 81-87, 2013

[7] M. Faiz, A. Zahari, K. Awanga and H. Hussin, "Corrosion inhibition on mild steel in 1 M HCl solution by Cryptocarya nigra extracts and three of its constituents (alkaloids)," *RSC Adv,* vol. 10 , pp. 6547–6562, 2020

[8] H. E. Fathabadia. M, Ghorbanib, H. Mokarami Ghartavolb," Corrosion Inhibition of Mild Steel with Tolyltriazole," *Materials Research*, vol.24 no. 4, 2021

[9] G. Karthik, M. Sundaravadivelu," Studies on the inhibition of mild steel corrosion in hydrochloric acid solution by atenolol drug," *Egyptian Journal of Petroleum*, 2015.

[10] A. A. Farag , A. S. Ismail, M.A. Migahed, "Environmental-friendly shrimp waste protein corrosion inhibitor for carbon steel in 1 M HCl solution," *Egyptian Journal of Petroleum,* 2018

[11] N.Gunavathy and S.C. Murugavel, "Corrosion Inhibition Studies of Mild Steel in Acid Medium Using Musa Acuminata Fruit Peel Extract," *E-Journal of Chemistry,* vol. 9 no.1, pp. 487-495, 2012

[12] Taleb H. Ibrahim, and Mohamed Abou Zour, "Corrosion Inhibition of Mild Steel using Fig Leaves Extract in Hydrochloric Acid Solution," *Int. J. Electrochem. Sci.*, vol.6, pp.6442 – 6455, 2011

[13] Malarvizhi Manickam, Dheenadhayalan Sivakumar, Brindha,Thirumalairaj, and Mallika Jaganathan," Corrosion Inhibition of Mild Steel in 1 mol L^{-1} HCl Using Gum Exudates of Azadirachta indica," *Hindawi Publishing Corporation Advances in Physical Chemistry*, Vol. 2016, Article ID 5987528, 12 pages.

[14] Alfred I. Onen , B. T. Nwufo , Eno E. Ebenso3, Reuben M. Hlophe3, Titanium (IV) Oxide as Corrosion Inhibitor for Aluminium and Mild Steel in Acidic Medium, *Int. J. Electrochem. Sci.,* vol. 5, pp.1563 – 1573, 2010

[15] Rupesh Kushwah, R. K. Pathak, "Inhibition of Mild Steel Corrosion in 0.5 M Sulphuric Acid Solution by Aspirin Drug," *International Journal of Emerging Technology and Advanced Engineering,* Vol. 4, no. 7, July 2014.

[16] W. Niouri1, B. Zerga, M. Sfaira, M. Taleb, B. Hammouti, M. Ebn Touhami, S.S. Al- Deyab H. Benzeid, El M. Essassi,"Substitution Effect of two Oxygen Atoms by Sulphur Atoms in New Synthesized Benzodiazepine Molecules towards Mild Steel Corrosion Inhibition in Hydrochloric Acid*," Int. J. Electrochem. Sci.,* vol. 7, pp.10190 – 10204, 2012

[17] M. A. Quraishi1, Sudheer, Eno E. Ebenso, "Ketorol: New and Effective Corrosion Inhibitor for Mild Steel in Hydrochloric Acid Solution",*Int. J. Electrochem. Sci.*, vol. 7, pp. 9920 – 9932. 2012

[18] Sutiana Junaedi, Ahmed A. Al-Amiery Abdulhadi Kadihm, dul Amir H. dhum and Abu Bakar Mohamad, " Inhibition Effects of a Synthesized Novel 4-Aminoantipyrine Derivative on the Corrosion of Mild Steel in Hydrochloric Acid Solution together with Quantum Chemical Studies*", Int. J. Mol. Sci.* vol. *14, no.*6, pp.11915-11928, 2013

[19] N. O. Eddy , B. I. Ita , E. E. Ebenso," Experimental and Theoretical Studies on the Corrosion Inhibition Potentials of some Anisole derivatives for Mild Steel", *Int. J. Electrochem.* Sci., vol.6, pp. 2101 – 2121, 2011

[20] I.B. Obot, E.E. Ebenso, Zuhair. M. Gasem, "Eco-friendly Corrosion Inhibitors: Adsorption and Inhibitive Action of Ethanol Extracts of Chlomolaena Odorata L. for the Corrosion of Mild Steel in H_2SO_4 Solutions", *Int. J. Electrochem. Sci.,* vol.7, pp.1997-2008, 2012

[22] S. A. Umoren, Z. M. Gasem, and I. B. Obot, "Natural Products for Material Protection :Inhibition of Mild Steel Corrosion by Date Palm Seed Extracts in Acidic Media*," Ind. Eng. Chem. Res*., vol. 52, pp. 14855–14865, 2013.

[23] I. B. Obot and N. O. Obi-Egbedi, "Anti-corrosive properties of xanthone on mild steel corrosion in sulphuric acid: Experimental and theoretical investigations," *Curr. Appl. Phys.,* vol. 11, no. 3, pp. 382–392, 2011

[24] G.Y.Elewady, I.A.El-said and A.S.Fouda,"Anion Surfactants as corrosion inhibitors for Aluminium dissolution in HCl solutions", *Int.J.Electrochem.Sci.,*vol. 3,pp.177-190, 2008

[25] D.J. Latha,Unnatii J.Naik, N.K.Shah, "Investigation of cumin extract as an eco-friendly green corrosion inhibitor for pure aluminium in acid medium", *journal of Materials and Environmental science"*.,vol.4, no.5, pp.701-708**,** 2013.

[26] Mahendra Yadav, Debasis Behera, Sumit Kumar, and Rajesh Ranjan Sinha, "Experimental and Quantum Chemical Studies on the Corrosion Inhibition Performance of Benzimidazole Derivatives for Mild Steel in HCl", *Ind. Eng. Chem.* 2013.

[27] Lukman O. Olasunkanmi, Ime B. Obot, Mwadham M. Kabanda and Eno E. Ebenso, "Some Quinoxalin-6-yl Derivatives as Corrosion Inhibitors for Mild Steel in Hydrochloric Acid: Experimental and Theoretical Studies", *The Journal of Physical Chemistry*, 2015.

[28] Priyanka Singh, Eno E. Ebenso,,Lukman O. Olasunkanmi,I. B. Obot and M.A. Quraishi, "Electrochemical, Theoretical, and Surface Morphological Studies of Corrosion Inhibition Effect of Green Naphthyridine Derivatives on Mild Steel in Hydrochloric Acid", *J. Phys. Chem.* Vol.120, pp. 3408–3419, 2016

[29] Jiyaul Haque, Chandrabhan Verma, Vandana Srivastava, and M. A. Quraishi, Eno E. Ebenso, "Experimental and quantum chemical studies of functionalized tetrahydropyridines as corrosion inhibitors for mild steel in 1 M Hydrochloric acid", *Results in Physics*, 2018

[30] Hong Ju, Li Ding, Can Sun, and Jie-jing Chen, "Quantum Chemical Study on the Corrosion Inhibition of Some Oxadiazoles", *Hindawi Publishing Corporation Advances in Materials Science and Engineering*, vol. 2015, Article ID 519606, 5 pages.

[31] Iman Danaee S. RameshKumar M. RashvandAvei M. Vijayan,"Electrochemical and Quantum Chemical Studies on Corrosion Inhibition Performance of 2,2'-(2-Hydroxyethylimino) bis[N-(alphaalpha-dimethylphenethyl)-N-methylacetamide] on Mild Steel Corrosion in 1M HCl Solution", *Materials Research*, vol.23, no.2, e20180610, 2020

[32] Kalaiselvi Kathirvel, Brindha Thirumalairaj, Mallika Jaganathan,"Quantum Chemical Studies on the Corrosion Inhibition of Mild Steel by Piperidin-4-One Derivatives in 1 M H_3PO_4", *Open Journal of Metal,* vol.4, pp.73-85, 2014

[33] Eno E. Ebenso , David A. Isabirye and Nnabuk O. Eddy , Adsorption and Quantum Chemical Studies on the Inhibition Potentials of Some Thiosemicarbazides for the Corrosion of Mild Steel in Acidic Medium, *Int. J. Mol. Sci.* vol.11,pp.2473-2498, 2010

[34] F. El-Taib Heakal, M.M. Osman, M.A. Deyab and A.E. Elkholy, "Electrochemical and Quantum Chemical Studies on the Corrosion Inhibition Potential of Camellia sinensis Leaves Extract for Carbon Steel in Produced Water", *Z. Phys. Chem.* 2017.

[35] Ahmed Al-Amiery1, Taghried A. Salman, Khalida F. Alazawi, Lina M. Shaker Abdul Amir H. Kadhum and Mohd S. Tariff, "Quantum chemical elucidation on corrosion inhibition efficiency of Schiff base: DFT investigations supported by weight loss and SEM techniques", *International Journal of Low-Carbon Technologies*, vol. 15, pp. 202–209, 2020

[36] Quy Huong Dinh, Tran Duong, and Nam Pham Cam, "Study of 1-Benzyl-3-phenyl-2-thiourea as an Effective Steel Corrosion Inhibitor in 1.0M HCl Solution", *Hindawi Journal of Chemistry,* vol. 2021, Article ID 5519411, 14 pages.

[37] M. Ouakki , M. Galai , M. Rbaa , A.S. Abousalem B. Lakhrissi , E.H. Rifif, M. Cherkaoui, "Quantum chemical and experimental evaluation of the inhibitory action of two imidazole derivatives on mild steel corrosion in sulphuric acid medium", *Heliyon*, vol.5, e02759,2019

[38] Ahmed Y. Musa, Abu Bakar Mohamad, Mohd Sobri Takriff, Ramzi T. T. Jalgham, "Electrochemical and quantum chemical studies on phthalhydrazide as corrosion inhibitor for mild steel in 1 M HCl solution", *Res Chem Intermed*, vol.38,pp.453—461, 2012

[39] Nnabuk O. Eddy, Femi E. Awe , Abdulfatai A. Siaka , Ladan Magaji , Eno E. Ebenso, "Chemical Information From GC-MS Studies of Ethanol Extract of Andrographis Paniculata and Their Corrosion Inhibition Potentials on Mild Steel in HCl Solution", *Int. J. Electrochem.* Sci., vol. 6, pp. 4316 – 4328, 2011

[40] Mwadham M Kabanda, Lutendo C. Murulana1, Muzaffer Ozcan, Faruk Karadag, Ilyas Dehri, I.B. Obot, Eno E. Ebenso1, "Quantum Chemical Studies on the Corrosion Inhibition of Mild Steel by Some Triazoles and Benzimidazole Derivatives in Acidic Medium", *Int. J. Electrochem. Sci.,* vol.7, pp. 5035 – 5056, 2012

[41] Nnabuk O. Eddy, Benedict I. Ita, Nkechi E. Ibisi, Eno E. Ebenso, "Experimental and Quantum Chemical Studies on the Corrosion Inhibition Potentials of 2-(2-Oxoindolin-3-Ylideneamino) Acetic Acid and Indoline-2,3-DioneExperimental and Quantum Chemical Studies on the Corrosion Inhibition Potentials of 2-(2-Oxoindolin-3-Ylideneamino) Acetic Acid and Indoline-2,3-Dione", *Int. J. Electrochem. Sci.,* vol. 6, pp.1027 – 104, 2011

[42] P.Udhayakala, T.V. Rajendiran, S. Gunasekaran, Journal of Advanced Scientific Research, "Quantum Chemical Studies on the Efficiencies of Vinyl Imidazole Derivatives as Corrosion Inhibitors For Mild Steel", *Journal of Advanced Scientific Research,* vol.3, no.2, pp. 37-44, 2012

[43] N. O. Eddy, B. I. Ita, E. E. Ebenso," Experimental and Theoretical Studies on the Corrosion Inhibition Potentials of some Anisole derivatives for Mild Steel", *Int. J. Electrochem. Sci.,* vol.6, pp. 2101 – 2121, 2011

[44] M.Abdallah, A.M. El Defrawy, I. A. Zaafarany, M.Sobhi, A.H.M. Elwahy, and M. R. Shaaban," Inhibition Effects and Theoretical Studies of Synthesized Novel Bisaminothiazole Derivatives as Corrosion Inhibitors for Carbon Steel in Sulphuric Acid Solutions", *Int. J. Electrochem. Sci.,* vol. 9, pp. 2186 – 2207, 2014

[45] Sutiana Junaedi, Ahmed A. Al-Amiery, Abdulhadi Kadihum, Abdul Amir H. Kadhumand Abu Bakar Mohamad," Inhibition Effects of a Synthesized Novel 4-Aminoantipyrine Derivative on the Corrosion of Mild Steel in Hydrochloric Acid Solution together with Quantum Chemical Studies", *Int. J. Mol. Sci.,*vol. 14, pp. 11915-11928, 2013

[46] M. M. Kabanda, S. K. Shukla, A. K. Singh, L. C. Murulana, and E. E. Ebenso, "Electrochemical and Quantum Chemical Studies on Calmagite and Fast Sulphone Black F dyes as Corrosion Inhibitors for Mild Steel in Hydrochloric Medium," *Int. J. Electrochem. Sci.*, vol. 7, pp. 8813–8831, 2012.

[47] M. Abdallah, A. M. El Defrawy, I. A. Zaafarany, M. Sobhi, and A. H. M. Elwahy, "Inhibition Effects and Theoretical Studies of Synthesized Novel Bisaminothiazole Derivatives as Corrosion Inhibitors for Carbon Steel in Sulphuric Acid Solutions," *Int. J. Electrochem. Sci.*, vol. 9, pp. 2186–2207, 2014.

[49] I. B. Obot and Z. M. Gasem, "Theoretical evaluation of corrosion inhibition performance of some pyrazine derivatives," *Corros. Sci.*, vol. 83, pp. 359–366, 2014.

[50] A. Aouniti, K. F. Khaled, and B. Hammouti, "Correlation Between Inhibition Efficiency and Chemical Structure of Some Amino Acids on the Corrosion of Armco Iron in Molar HCl," *Int. J. Electrochem. Sci.,* vol. 8, pp. 5925–5943, 2013.

[51] S. Junaedi, A. a Al-Amiery, A. Kadihum, A. A. H. Kadhum, and A. B. Mohamad, "Inhibition Effects of a Synthesized Novel 4-Aminoantipyrine Derivative on the Corrosion of Mild Steel in Hydrochloric Acid Solution together with Quantum Chemical Studies.," *Int. J. Mol. Sci.*, vol. 14, no. 6, pp. 11915–28, 2013.

[52] N. O. Obi-Egbedi and I. B. Obot, "Xanthione: A new and effective corrosion inhibitor for mild steel in sulphuric acid solution," *Arab. J. Chem.*, vol. 6, no.2, pp. 211–223, 2013.

[53] H. Zarrok, H. Oudda, A. Zarrouk, R. Salghi, B. Hammouti, and M. Bouachrine, "Weight Loss Measurement and Theoretical Study of New Pyridazine Compound as Corrosion Inhibitor for C38 Steel in Hydrochloric Acid Solution," *Der Pharma Chem.*, vol. 3, no. 6, pp. 576–590, 2011.

[54] S. John and A. Joseph, "Theoretical and electrochemical studies on the effect of substitution on 1, 2, 4- triazole towards mild steel corrosion inhibition in hydrochloric acid," *Int. J. Chem. Technol.*, vol. 19, pp. 195–204, 2012.

[55] F. Bentiss and M. Lagrenée, "Heterocyclic compounds as corrosion inhibitors for mild steel in hydrochloric acid medium — correlation between electronic structure and inhibition efficiency," *J. Mater. Environ. Sci.*, vol. 2, no. 1, pp. 13–17, 2011.

[56] P. Liu, X. Fang, Y. Tang, C. Sun, and C. Yao, "Electrochemical and Quantum Chemical Studies of 5-Substituted Tetrazoles as Corrosion Inhibitors for Copper in Aerated 0.5 M H_2SO_4 Solution," *Mater. Sci. Appl.*, vol. 02, no. 09, pp. 1268–1278, 2011.

[57] G. Gece and S. Bilgiç, "Quantum chemical study of some cyclic nitrogen compounds as corrosion inhibitors of steel in NaCl media," *Corros. Sci.*, vol. 51, no. 8, pp. 1876–1878, 2009.

[58] K. F. Khaled and M. A. Amin, "Dry and wet lab studies for some benzotriazole derivatives as possible corrosion inhibitors for copper in 1 . 0 M HNO_3," *Corros. Sci.*, vol. 51, no. 9, pp. 2098–2106, 2009.

[59] K. F. Khaled, "Molecular simulation, quantum chemical calculations and electrochemical studies for inhibition of mild steel by triazoles," *Electrochim. Acta*, vol. 53, no. 9, pp. 3484–3492, 2008.

[60] L. Larabi, Y. Harek, O. Benali, and S. Ghalem, "Hydrazide derivatives as corrosion inhibitors for mild steel in 1M HCl," *Prog. Org. Coatings*, vol. 54, no. 3, pp. 256–262, 2005.

[61] H. R. Obayes, G. H. Alwan, A. H. M. Alobaidy, A. Al-Amiery, A. A. H. Kadhum, and A. B. Mohamad, "Quantum chemical assessment of benzimidazole derivatives as corrosion inhibitors.," *Chem. Cent. J.*, vol. 8, no. 1, pp. 21, 2014.

[62] Nnabuk O. Eddy, Momoh-Yahaya H, Emeka E. Oguzie,"Theoretical and experimental studies on the corrosion inhibition potentials of some purines for aluminum in 0.1 M HCl", *Journal of Advanced Research*, vol. 6, pp. 203-217, 2015.

[63] Majid Khodaei-Tehrani and Ali Niazl, "Quantum Chemical Studies on the Corrosion Inhibition of Some Hector Bases on Mild Steel in Acidic Medium", *Oriental. J. of chemistry*, vol. 31, no.1, pp.423-429, 2015

[64] N. O. Eddy and E. E. Ebenso, "Adsorption and Quantum Chemical Studies on Cloxacillin and Halides for the Corrosion of Mild Steel in Acidic Medium," *Int. J. Electrochem. Sci.*, vol. 5, pp. 731–750, 2010.

[65] M. M. Kabanda, S. K. Shukla, A. K. Singh, L. C. Murulana, and E. E. Ebenso, "Electrochemical and Quantum Chemical Studies on Calmagite and Fast Sulphone Black F dyes as Corrosion Inhibitors for Mild Steel in Hydrochloric Medium," *Int. J. Electrochem. Sci.*, vol. 7, pp. 8813–8831, 2012.

[66] I.B. Obot1, N.O. Obi-Egbedi S.A. Umoren1, Adsorption Characteristics and Corrosion Inhibitive Properties of Clotrimazole for Aluminium Corrosion in Hydrochloric Acid, *Int. J. Electrochem. Sci.,* 4, pp. 863 – 877, 2009

[67] Ambrish Singh, Eno. E. Ebenso, M. A. Quraishi," Theoretical and Electrochemical Studies of Metformin as Corrosion Inhibitor for Mild Steel in Hydrochloric Acid Solution", *Int. J. Electrochem. Sci*., 7, pp. 4766 – 4779, 2012

[68] Salah Eid," Expired Desloratidine Drug as Inhibitor for Corrosion of Carbon Steel Pipeline in Hydrochloric acid Solution", *Int. J. Electrochem. Sci.,* vol.16, pp.150852, 2021

[69] A. S. Fouda, M.N. EL-Haddad and Y.M. Abdallah, "Septazole: Antibacterial Drug as a Green Corrosion Inhibitor for Copper in Hydrochloric Acid Solutions", *International Journal of Innovative Research in Science, Engineering and Technology*, vol. 2, no. 12, 2013

[70] A. S. Fouda, A. M. El-Defrawy, and M. W. El-Sherbeni, Lornoxicam & Tenoxicam, "Drugs as Green Corrosion Inhibitors for Carbon Steel in 1 M H2SO4 Solution", *Journal of Electrochemical Science and Technology*, vol. 4, no. 2, pp.47-56, 2013

[71] L. Magaji, P. O. Ameh, N. O. Eddy, A. Uzairu, A. A. Siaka, S. Habib, A. M. Ayuba 1, S. M. Gumel," Ciprofloxacin as Corrosion Inhibitors for Mild Steel -Effects of Concentration and Temperature", *International Journal of Modern Chemistry,* vol. 2, no.2, pp.64-73, 2012

[72] R. S. Abdel Hameed, H. I. AlShafey, A. H. Abu-Naw was "2,3, 2-(2, 6-dichloranilino) phenyl acetic acid Drugs as Eco-Friendly Corrosion Inhibitors for Mild Steel in 1M HCl", *Int. J. Electrochem. Sci.*, vol.9, pp.6006 – 6019, 2014

[73] Imran Naqvi, A. R. Saleemi, S. Naveed, "Cefixime: A drug as Efficient Corrosion Inhibitor for Mild Steel in Acidic Media. Electrochemical and Thermodynamic Studies". *Int. J. Electrochem. Sci.,* vol.6, pp. 146 – 161, 2011

[74] R.S. Abdel Hameed, "Ranitidine Drugs as Non-Toxic Corrosion Inhibitors for Mild Steel in Hydrochloric Acid Medium, Portugaliae Electrochimica Acta", vol. 29, No.4, 273-285, 2011

[75] I. A. Akpan, N. O. Offiong, "Electrochemical Study of Corrosion Inhibition Action of Sulphadoxine and Pyrimethamine on Mild Steel in Acidic Medium", *International Journal of Chemical and Process Engineering Research*, vol. 1, No. 2, pp. 10-18, 2014

[76] A. S. Fouda, H. S. El-Desoky H.S, Abdel-Galeil M.A, Dina Mansour, "Niclosamide and dichlorphenamide: New and effective corrosion inhibitors for carbon steel in 1M HCl solution". *SN Applied Sciences*, vol.3, pp. 287, 2021

[77] Omar A. Hazazi, and M.Abdallah,"1, 2, Prazole Compounds as Inhibitors for Corrosion of Aluminum in Hydrochloric acid", *Int. J. Electrochem. Sci.*, vol. 8, pp.8138 – 8152, 2013

[78] Nicolae Vaszilcsin· Valentin Ordodi, Alexandra Borza, "Corrosion inhibitors from expired drugs", *Int J Pharm,* vol. 431, no. 1-2, 2012

[79] Gökhan Gece," Drugs: A review of promising novel corrosion inhibitors", *Corrosion Science*, vol. 53, pp. 3873, 2011

[80] Ashish Kumar Singh, Eno E. Ebenso, Effect of Ceftezole on the Corrosion of Mild Steel in HCl Solution, *Int. J. Electrochem. Sci.*, vol.7, pp. 2349 – 2360, 2012

[81] Ahmad M. El-Desoky, Hytham M. Ahmed and Alaa E. Ali, Electrochemical and Analytical Study of the Corrosion Inhibitory behavior of Expired Pharmaceutical Compounds for C- Steel Corrosion, *Int. J. Electrochem. Sci.*, vol.10, pp.5112 – 5129, 2015

[82] R. S. Abdel Hameed, E. A. Ismail, A. H. Abu- Nawwas and Hussin I. AL-Shafey, Expired Voltaren Drugs as Corrosion Inhibitor for Aluminium in Hydrochloric Acid, *Int. J. Electrochem. Sci.*, vol. 10, pp. 2098 – 2109, 2015.

[83] Narayana Hebbar, BM Praveen, BM Prasanna and TV Venkatesha," Corrosion inhibition behavior of ketosulphide for mildsteel in acidic medium", International Research Journal of Chemistry, vol .2, no.1, pp.018-020, 2015.

[84] Nadia Hammouda, Hacène Chadli, Gildas Guillemot, Kamel Belmokre," The corrosion Protection Behaviour of Zinc,Rich Epoxy Paint in 3% NaCl Solution", *Advances in Chemical Engineering and Science,* vol.1, pp.51-60, 2011

[85] B.M. Durodola, J.A.O. Olugbuyiro, S.A. Moshood, O.S. Fayomi . A.P.I. Popoola,"Study of Influence of Zinc Plated Mild Steel Deterioration in Seawater Environment", *Int. J. Electrochem. Sci.,* vol. 6, pp. 5605 – 5616, 2011

[86] Muhammad Firdaus Mohd Nazeri, Mohd Shahadan Mohd Suan, Mohamad Najmi Masri, Nurhaswani Alias, and Ahmad Azmin Mohamad, "Corrosion Studies of Conductive Paint Coating Using Battery,Cathode Waste Material in Sodium Chloride Solution", *Int. J. Electrochem. Sci.*, vol.7, pp. 6976 – 6987, 2012

[87] K. L. Palanisamy1, V. Devabharathi And N. Meenakshi Sundaram," Corrosion Inhibition Studies Of Mild Steel With Carrier Oil Stabilized Of Iron Oxide Nanoparticles Incorporated Into A Paint ", *International Journal of ChemTech Research*,vol.7, no.4, pp.1661-1664, 2014-2015

[88] Rosa Vera1, Judith Apablaza, Ana M. Carvajal, Enrique Vera," Effect of Surface Coatings in the Corrosion of Reinforced Concrete in Acid Environments", *Int. J. Electrochem. Sci.* vol.8, pp.11832 – 11846, 2013

[89] Cook A, Gabriel A, Laycock N, " On the Mechanism of Corrosion Protection of Mild Steel with Polyaniline, Journal of the Electrochemical Society",vol.151, no.9, pp.B529-B535, 2004

[90] Adriana Samide, Gabriela Eugenia Iacobescu, Bogdan Tutunaru, Roxana Grecu, Cristian Tigae and Cezar Spînu, "Inhibitory Properties of Neomycin thin Film Formed on Carbon Steel in Sulfuric Acid Solution: Electrochemical and AFM Investigation", *Coatings*, vol.7, pp.181, 2017

[91] Chandrabhan Verma, D.S. Chauhan, M.A. Quraishi, "Drugs as environmentally benign corrosion inhibitors for ferrous and nonferrous materials in acid environment: An overview", *Journal of Materials and Environmental Sciences,* vol. 8, no. 11, pp.4040-4051, 2017

[92] M. Menaga, D. Nalini and Sowmya Ramkumar, "Experimental Results with Hypothetical Evidences for Bexol as an Effective Inhibitor on Corrosion Inhibition of Mild Steel in Acidic Environment and its Epoxy Coating Behavior", *Orient. J. Chem.,* vol. 36, no 3, pp. 504-512, 2020.

[93] Lekan Taofeek Popoola," Progress on pharmaceutical drugs, plant extracts and ionic liquids as corrosion inhibitors", *Heliyon*, vol.5, e01143, 2019

[94] Xihua Xu, Ambrish Singh, Zhipeng Sun, K. R. Ansari and Yuanhua Lin, "Theoretical, thermodynamic and electrochemical analysis of biotin drug as an impending corrosion inhibitor for mild steel in 15% hydrochloric acid", *Royal society open science*, vol.4, pp.170933.

[95] S.A. Umoren and M.M. Solomon," Recent Developments on the Use of Polymers as Corrosion Inhibitors - A Review", *The Open Materials Science Journal*, vol.8, pp.39-54, 2014

[96] Pang xuehui, Gong Min, Zhang yuxuan, Wei qin & Hou baorong," Corrosion inhibition and mechanism of mild steel in hydrochloric acid by ceftriaxone and amoxicillin", *Sci China Chem*, vol.54 No.9, 2011

[97] Mene Benmahammed, Tahar Douadi, Saifi Issaadi, Djamel Daoud, Salah Chafaa,"Electrochemical, DFT and MD Simulation Investigations of the Corrosion Inhibition Properties of Heterocyclic Sulfide Derivatives for Mild Steel in Acidic Medium", *Surface Review and Letters*, 2019.

[98] Akhil Saxena, Dwarika Prasad and Rajesh Haldhar, "Use of Syzygium aromaticum extract as Green Corrosion Inhibitor for Mild Steel in 0.5 M H_2SO_4", *Surface Review and Letters*, 2018.

[99] Parul Dohare, D.S. Chauhan, A.A. Sorour, M.A. Quraishi," DFT and experimental studies on the inhibition potentials of expired Tramadol drug on mild steel corrosion in hydrochloric acid", *Materials Discovery*, 2017.

Chapter III

CHAPTER III

MATERIALS AND METHODS

3.1. Selection of Materials

3.1.1. Selection of the Inhibitor

A survey of Literature reveals that the inhibition efficiency of the selected inhibitors Bexol, Monit, Cilny, Prazopill have not reported so far and this makes the present investigation in context of the priorities to select pharmaceutical expired drugs, inhibitor system as mentioned in the **Table 3.1.**

Table 3.1. Inhibitor system used in the current investigation

S. No	Structure of the compound	Name of the compound
1.	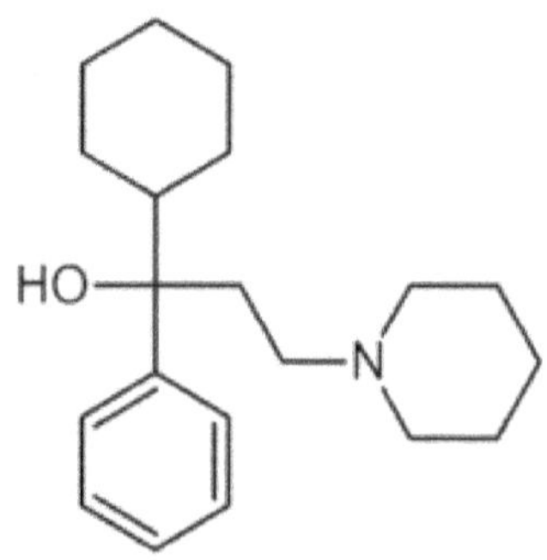 (RS)-1-Cyclohexyl-1-phenyl-3-(1-piperidyl)propan-1-ol	BEXOL-DT
2.	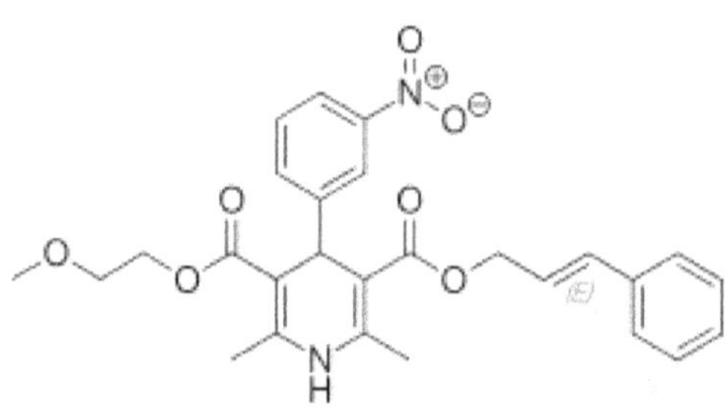 3-(E)-3-Phenyl-2-propenyl 5-2-methoxyethyl 2,6-dimethyl-4-(m-nitrophenyl)-1,4-dihydropyridine-3,5-dicarboxylate	CILNY

3.	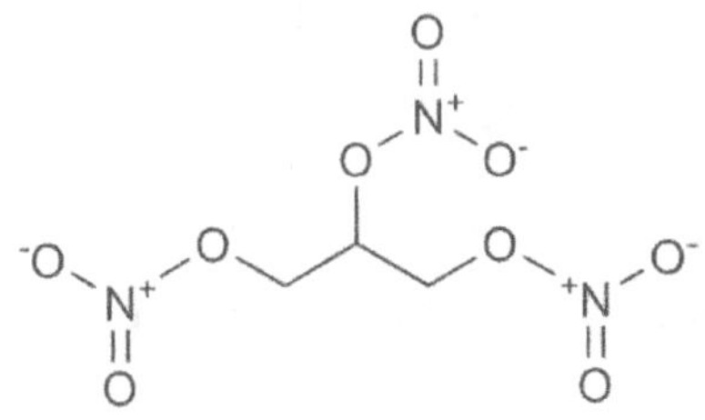 Propane-1,2,3-triyl trinitrate	MONIT
4.	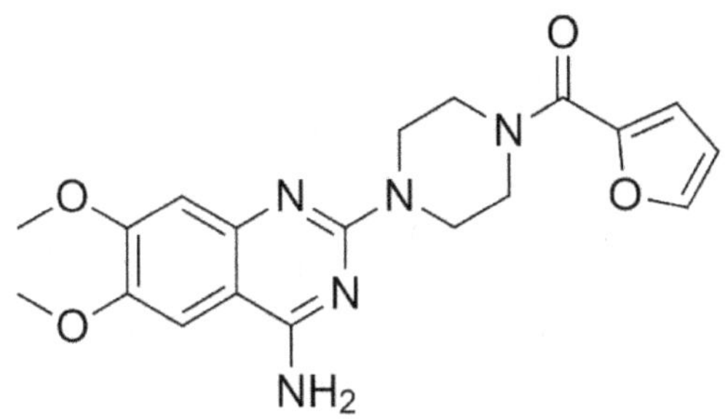 [4-(4-Amino-6,7-dimethoxy-2-quinazolinyl)-1-piperazinyl](2-furyl)methanone	PRAZOPILL

1. BEXOL-DT

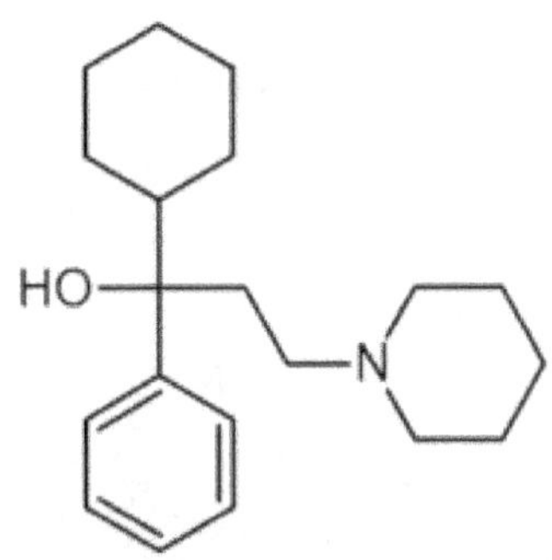

The selected properties are listed below:

Molecular mass	: 301.474 g·mol^{-1}
Molecular formula	: $C_{20}H_{31}NO$
Trade name	: Artane, Parkin
Manufacturer	: Intas pharmaceutical Ltd., Gujarath

Medicinal use : This improves muscle control and reduces stiffness in Parkinson's disease. It also improves movement disorder (restlessness, involuntary movements or muscle spasm) caused by certain other medicines.

2. CILNY

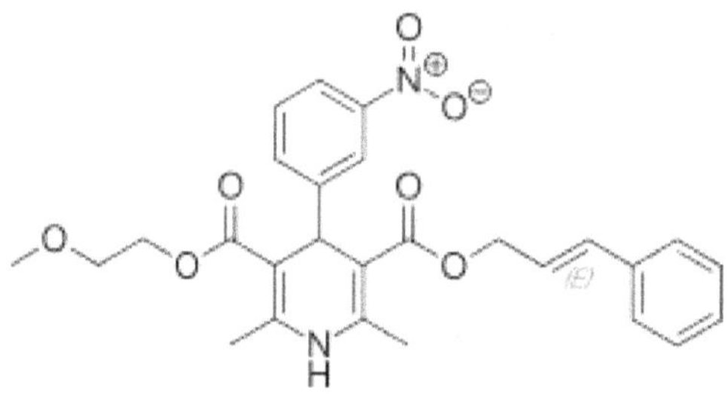

The selected properties are listed below:

Molecular mass : 492.52 g/mol

Molecular formula : $C_{27}H_{28}N_2O_7$

Trade name : Atelec

Manufacturer : Intas pharmaceutical Ltd., Gujarath

Medicinal use : It is used to treat high blood pressure (hypertension

C. MONIT

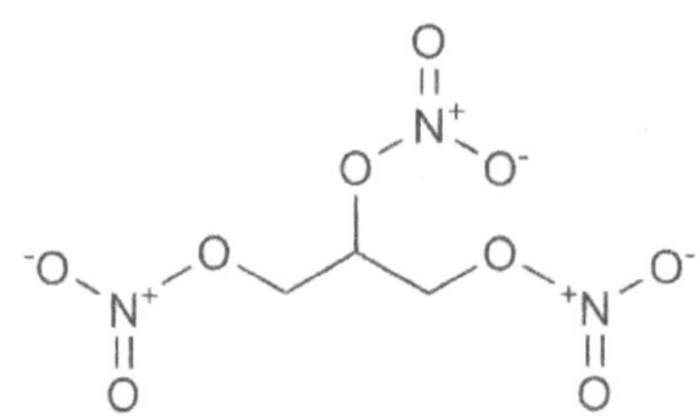

The selected properties are listed below:

Molecular mass : 227.085 $g{\cdot}mol^{-1}$

Molecular formula : $C_3H_5N_3O_9$

Trade name : Nitrol,

Manufacturer : Intas pharmaceutical Ltd., Gujarath

Medicinal use : Treatment of Angina (heart-related chest pain)

4. PRAZOPILL

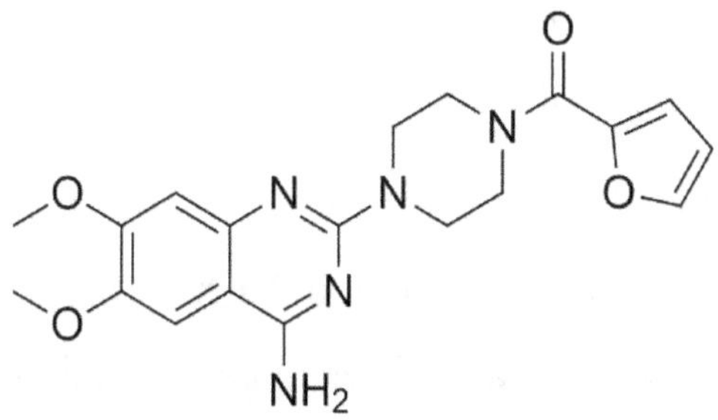

The selected properties are listed below:

Molecular mass : 383.401 g/mol

Molecular formula : $C_{19}H_{21}N_5O_4$

Trade name : Minipress, Vasoflex,

Manufacturer : Intas pharmaceutical Ltd., Gujarath

Medicinal use : Hypertension (high blood pressure)

3.1.2. Selection of the Metal

Mild steel is the Metal used in almost all industries like food, petroleum, Power production, Chemical industries and domestic purposes due to its high tensile strength and its mechanical properties [1]–[6]. This metal undergoes deterioration when exposed to acidic medium [7]–[10].

i. Preparation of test coupons

The mild steel having a composition (wt. %) of 0.098 % C, 0.201 % Mn, 0.02% P,0.016% S,0.012% Ni, and 99.653% Fe, were cut into the dimension 5.00 cm × 1.50 cm × 0.50 cm for weight loss measurements. For electrochemical measurements, mild steel rod of exposed surface area 1cm^2 while the remainder was embedded by Teflon with 3cm long stem following ASTM G 102-89 procedure was employed. Prior to each experiment, the specimens were abraded consecutively with emery papers from 400 to 1000 grit, then washed with distilled water, degreased with acetone, finally dried at room temperature.

3.1.3. Preparation of acid solution

The corrosive medium 1 M HCl was prepared by using analytical grade hydrochloric acid. The 1M HCl solution was studied for further studies.

3.1.4. Preparation of Inhibitor

The testing stock solution of 100 mg/L of BEX, CIL, MON and PYR was prepared by dissolving it in 1M HCl solution and different concentrations (BEX : 5 – 50 mg/L) of the inhibitor solution are prepared by dilution with double distilled water.

3.2. Methods adopted

3.2.1 Weight loss measurements

Rectangular coupons of working surface area 5 cm^2 were used for weight loss measurements following the standard ASTM G1-03 procedure. Cleaned and weighed mild steel samples in triplicate were immersed in 1M HCl solution with and without different concentrations of BEX, CIL, MON, PYR for 3 h at 298 K, respectively. Then the samples were taken out, scrubbed with a bristle brush, cleaned by distilled water and acetone, then dried and reweighed [11-13]. From the weight loss measurements the surface coverage (θ), inhibition efficiency (IE %) and corrosion rate (mm/y) at different concentrations are calculated as follows and listed in Table 1.

$$IE\% = \frac{W_o - W_i}{W_o} * 100 \quad (1)$$

where W_o and W_i are the weight loss of the mild steel coupon without and with the inhibitor respectively. Corrosion rate (mm/y) is calculated from the following equation,

$$Corrosion\ rate(mm/y) = \frac{87.6*10^{3}*W}{D*A*T} * 100 \quad (2)$$

where, W –Weight loss in milligrams (mg), D –Density of specimen in grams per cubic centimeter (g/cm^3), A - surface area of specimen in square centimeter (cm^2) and T – exposure time in hour (h).

3.2.2 Electrochemical measurements

Electrochemical measurements were carried out in a traditional three-electrode cell by a three electrode electrochemical workstation. Mild steel rod of exposed surface area 1cm^2 was used as a working electrode. Saturated calomel electrode (SCE) and Platinum electrode were treated as reference and counter electrodes, respectively. All potentials in the present study were measured with respect to SCE. Prior to each measurement, the working electrode was immersed in the aggressive media for 30 min until a steady state of open circuit potential (OCP) was obtained [14 - 16]. Subsequently, EIS measurements were carried out at the OCP over a frequency range of 10 mHz to 100 kHz with a sinusoidal AC perturbation of 25 mV peak-to-peak [17 - 18]. Polarization curves were recorded at a scan rate of 100 mv s^{-1} in the potential range of -200 mv to +200 mv versus the OCP. The I.E % values are calculated from the following equation,

$$I.E = \frac{R_{ct(p)} - R_{ct(a)}}{R_{ct(p)}} * 100 \qquad (3)$$

The values of IE% is calculated as follows

$$IE\ (\%) = \frac{I_{corr(B)} - I_{corr(I)}}{I_{corr(B)}} * 100 \qquad (4)$$

where $R_{ct(p)}$ is the charge transfer resistance values of presence of inhibitor and $R_{ct(a)}$ is the charge transfer values of absence of inhibitor and $I_{corr(B)}$ and $I_{corr(I)}$are the uninhibited and inhibited corrosion current densities of the mild steel in 1M HCl.

3.2.3. Preparation of Epoxy Coating

Materials required for preparing 5ml of paint

Zinc -2.675g	Zinc phosphate – 1g
TiO_2-0.4287g	Volume of binder-1.25ml
Volume of solvent-3 ml	Inhibitor-0.1g

The fine powders of zinc, zinc phosphate, TiO_2 and the studied inhibitor powders were mixed well. Epoxy resin (binder) is mixed with polyamide (hardener). Xylene is used a solvent. The prepared fine powder is added with epoxy resin-polyamide mixture and grain well and solvent is added drop by drop until the paint form was obtained. The obtained paint is coated on the mild steel surface. Then they were labelled as (A) the bare mild steel (B) paint coated mild steel and (C) paint mixed with the inhibitor coated mild steel. The coated mild steel plates were dried in the laboratory at room temperature for 78 hrs. The epoxy coated mild steel with a thickness of 2mm coating were immersed in 3.5% of sodium chloride solution.

3.2.4. Electrochemical Measurements of the epoxy coating

Electrochemical measurements were performed using electrochemical system and electrochemical cell assembly of three electrodes. Coated mild steel coupons is a working electrode in area of $5cm^2$, calomel electrode as reference electrode and Platinum electrode as counter electrode, immersed in 3.5% of NaCl solution [19 - 20].

3.3. Theoretical methods

Complete optimized geometrical structure of the investigated inhibitor was obtained using density functional theory (DFT) with (B3LYP) procedure [21]–[23] within the basic set of 6-31G (d,p) within the program package. All the quantum chemical parameters such as the energies of the highest occupied Molecular orbital (E_{HOMO}) and lowest unoccupied molecular orbitals(E_{LUMO}), electronegativity(χ), dipole moment(μ), global hardness (η) and softness (σ), energy gap (ΔE), fraction of electrons transferred (ΔN) were calculated and discussed for the studied compound.

i. Global parameters - At the minimum point of potential energy surface for the optimized structure, molecular parameters such as bond length and bond angle, frontier molecular orbital energy (E_{HOMO} & E_{LUMO}), energy gap (ΔE), dipole moment (μ), electronegativity (χ), global hardness (η), softness (σ), the fraction of electron transferred (ΔN) from inhibitors to metal, Mulliken charge density, Fukui indices (*f*), condensed softness indices (S) are described [24]–[27]. These quantities are often defined following Koopmans' theorem [28], [29]. The energy gap can be determined by the equation:

$$\Delta E = E_{LUMO} - E_{HOMO} \quad (1)$$

Also, the ionization energy (IP) and the electron affinity (EA) of the molecules were calculated using the ground state energies of the respective systems as follows:

$$IP = -E_{HOMO};\ EA = -E_{LUMO} \quad (2)$$

Electronegativity, hardness and softness have proved to be very useful quantities in the chemical reactivity theory and they have been calculated as:

$$\chi = \frac{IP+EA}{2} \quad (3)$$

$$\eta = \frac{IP-EA}{2} \qquad 4$$

$$\sigma = \frac{1}{\eta} \qquad 5$$

For a reaction of two systems with different electronegativity (as a metallic surface and an inhibitor molecule) the following mechanism will take place: the electronic flow will occur from the molecule with the lower electro negativity toward that of higher value, until the chemical potentials are the same. The difference in electronegativity drives the electron transfer and the sum of the hardness parameters acts as a resistance.

The global Electrophilicity index was introduced by Parr and this index measures the tendency of chemical species to accept electrons [30]

$$\omega = \frac{\mu^2}{2\eta} \qquad 6$$

The fraction of transferred electrons (ΔN) from the inhibitor molecule to the metallic atom was calculated according to Pearson.

$$\Delta N = \frac{\chi_{Fe}-\chi_{Inh}}{2(\eta_{Fe}+\eta_{Inh})} \qquad 7$$

where χFe and χinh denote the absolute electronegativity of iron and inhibitor molecule respectively η_{Fe} and η_{inh} denote the absolute hardness of iron and the inhibitor molecule respectively. Another useful quantity that was defined in hard and soft acid– base (HSAB) theory is energy change during charge transfer from inhibitor to metallic atoms [30], [31]. This quantity is defined as,

$$\Delta E = -\frac{1}{4} * \frac{(\chi_{Fe}-\chi_{Inh})^2}{(\eta_{Fe}+\eta_{Inh})} \qquad 8$$

ii. Local parameters - The local selectivity of a corrosion inhibitor is best analysed by means of condensed Fukui function. The Fukui function is motivated by the fact that if an electron d is transferred to an N electron molecule, it will tend to distribute so as to minimize the energy of the resulting N + d electron system. The resulting change in electron density is the nucleophilic (f+) and electrophilic (f-) Fukui functions and can be calculated using the finite difference approximation as follows :

$$f^+ = q_{N+1} - q_N \text{ for nucleophilic attack} \qquad 9$$

$$f^- = q_N - q_{N-1} \text{ for electrophilic attack} \quad 10$$

Where, qN+1, qN and qN-1 are the density of electron is the Mulliken charge of the atom with N+1, N and N-1 electrons. Local softness indices allow the comparison of reactivity between similar atoms of different molecules, calculated easily starting from the relation between the Fukui function f and the condensed softness (σ) using the following equations

$$S^+ = (f^+)\sigma \quad 11$$

$$S^- = (f^-)\sigma \quad 12$$

Results pertaining to the present investigation are discussed in detail in the following chapters as tabulated below in **Table 3.2.**

Table 3.2: Synopsis of the present investigation

Chapter	Title of the Chapter	Synopsis
IV	**INVESTIGATION ON THE CORROSION INHIBITOR SYSTEM WITH WEIGHT LOSS MEASUREMENTS**	Deals with the studies on the inhibitive performance of the pharmaceutically active drugs BEX, CIL, MON and PYR using weight loss.
V	**EVALUATION OF CORROSION POTENTIAL OF THE STUDIED DRUGS BY ELECTROCHEMICAL METHODS**	Deals with the corrosion studies on the inhibitive system using polarization and impedance method
VI	**THEORETICAL INVESTIGATION ON THE INHIBITIVE PROPERTIES OF THE STUDIED DRUGS**	Deals with the theoretical studies on the investigated inhibitor systems for their corrosion protection ability
VII	**ELECTROCHEMICAL INVESTIGATION ON THE PROTECTIVE DRUG/EPOXY COMPOSITES IN 3.5 % NaCl**	Deals with evaluation of epoxy coated mild steel using EIS measurements at different time intervals.
VIII	**SUMMARY AND CONCLUSION**	Deals with the findings of the present investigation

References

[1] Y. Qiang, S. Zhang, L. Guo, X. Zheng, B. Xiang, and S. Chen, "Experimental and theoretical studies of four allyl imidazolium-based ionic liquids as green inhibitors for copper corrosion in sulfuric acid," *Corros. Sci.*, vol. 119, pp. 68–78, 2017.

[2] O. S. I. Fayomi, A. Sode, I. G. Akande, A. P. I. Popoola, and O. Agboola, "Improving the structural properties and corrosion behaviour of electroless deposited Ni-P-Zn coatings on mild steel for advanced processes," *AIMS Mater. Sci.*, vol. 7, no. 4, pp. 441–452, 2020.

[3] F. E. Abeng, V. C. Anadebe, P. Y. Nkom, K. J. Uwakwe, and E. G. Kamalu, "Experimental and theoretical study on the corrosion inhibitor potential of quinazoline derivative for mild steel in hydrochloric acid solution," *J. Electrochem. Sci. Eng.*, vol. 11, no. 1, pp. 11–26, 2021.

[4] X. Xu, A. Singh, Z. Sun, K. R. Ansari, and Y. Lin, "Theoretical, thermodynamic and electrochemical analysis of biotin drug as an impending corrosion inhibitor for mild steel in 15% hydrochloric acid," *R. Soc. Open Sci.*, vol. 4, no. 12, 2017.

[5] Z. Mirzakhanzadeh, A. Kosari, M. H. Moayed, R. Naderi, P. Taheri, and J. M. C. Mol, "Enhanced corrosion protection of mild steel by the synergetic effect of zinc aluminum polyphosphate and 2-mercaptobenzimidazole inhibitors incorporated in epoxy-polyamide coatings," *Corros. Sci.*, vol. 138, no. 2017, pp. 372–379, 2018.

[6] A. Fitoz, H. Nazır, M. Özgür (nee Yakut), E. Emregül, and K. C. Emregül, "An experimental and theoretical approach towards understanding the inhibitive behavior of a nitrile substituted coumarin compound as an effective acidic media inhibitor," *Corros. Sci.*, vol. 133, no. 2017, pp. 451–464, 2018.

[7] M. A. Aziz, Z. H. Khan, M. S. Khatun, and M. R. Hasan, "Corrosion Inhibition Study of Mild Steel in Acidic Medium by Antibiotic Drugs: A Comparative Study," *Aceh Int. J. Sci. Technol.*, vol. 3, no. 1, pp. 19–26, 2014.

[8] X. Luo, X. Pan, S. Yuan, S. Du, C. Zhang, and Y. Liu, "Corrosion inhibition of mild steel in simulated seawater solution by a green eco-friendly mixture of glucomannan (GL) and bisquaternary ammonium salt (BQAS)," *Corros. Sci.*, vol. 125, no. June, pp. 139–151, 2017.

[9] D. T. Oyekunle, O. Agboola, and A. O. Ayeni, “Corrosion Inhibitors as Building Evidence for Mild Steel: A Review,” in *Journal of Physics: Conference Series*, 2019, vol. 1378, no. 3.

[10] B. Nikpour, B. Ramezanzadeh, G. Bahlakeh, and M. Mahdavian, “Synthesis of graphene oxide nanosheets functionalized by green corrosion inhibitive compounds to fabricate a protective system,” *Corros. Sci.*, vol. 127, no. August, pp. 240–259, 2017.

[11] A. Nahlé *et al.*, “Novel triazole derivatives as ecological corrosion inhibitors for mild steel in 1.0 M HCl: experimental & theoretical approach,” *RSC Adv.*, vol. 11, no. 7, pp. 4147–4162, 2021.

[12] C. Verma, M. A. Quraishi, and A. Singh, “5-Substituted 1H-tetrazoles as effective corrosion inhibitors for mild steel in 1 M hydrochloric acid,” *J. Taibah Univ. Sci.*, vol. 10, no. 5, pp. 718–733, Sep. 2016.

[13] R. Ganapathi Sundaram and M. Sundaravadivelu, “Surface protection of mild steel in acidic chloride solution by 5-Nitro-8-Hydroxy Quinoline,” *Egypt. J. Pet.*, vol. 27, no. 1, pp. 95–103, 2018.

[14] M. Goyal, S. Kumar, I. Bahadur, C. Verma, and E. E. Ebenso, “Organic corrosion inhibitors for industrial cleaning of ferrous and non-ferrous metals in acidic solutions: A review,” *J. Mol. Liq.*, vol. 256, no. February, pp. 565–573, 2018.

[15] M. Mobin, S. Zehra, and R. Aslam, “L-Phenylalanine methyl ester hydrochloride as a green corrosion inhibitor for mild steel in hydrochloric acid solution and the effect of surfactant additive,” *RSC Adv.*, vol. 6, no. 7, pp. 5890–5902, 2016.

[16] F. AS, E. M. MA, and E.-M. T, “Investigation of the Inhibition of Carbon Steel Corrosion in Hydrochloric Acid Solutions by Domperidone Drug,” *J. Anal. Pharm. Res.*, vol. 5, no. 5, 2017.

[17] H. U. Nwankwo, L. O. Olasunkanmi, and E. E. Ebenso, “Experimental, quantum chemical and molecular dynamic simulations studies on the corrosion inhibition of mild steel by some carbazole derivatives,” *Sci. Rep.*, vol. 7, no. 1, 2017.

[18] B. Ramezanzadeh, S. Niroumandrad, A. Ahmadi, M. Mahdavian, and M. H. Mohamadzadeh Moghadam, "Enhancement of barrier and corrosion protection performance of an epoxy coating through wet transfer of amino functionalized graphene oxide," *Corros. Sci.*, vol. 103, pp. 283–304, 2016.

[19] K. Karattu Veedu, T. Peringattu Kalarikkal, N. Jayakumar, and N. K. Gopalan, "Anticorrosive Performance of Mangifera indica L. Leaf Extract-Based Hybrid Coating on Steel," *ACS Omega*, vol. 4, no. 6, pp. 10176–10184, 2019.

[20] S. C. Das, H. S. Pouya, and E. Ganjian, "Zinc-Rich Paint As Anode for Cathodic Protection of Steel in Concrete," *J. Mater. Civ. Eng.*, vol. 27, no. 11, p. 04015013, 2015.

[21] M. Mousavi and T. Baghgoli, "Application of interaction energy in quantitative structure-inhibition relationship study of some benzenethiol derivatives on copper corrosion," *Corros. Sci.*, vol. 105, pp. 170–176, 2016.

[22] S. Ramkumar and D. Nalini, "Enhanced corrosion resistance of mild steel in sulphuric acid medium by imidazole derivative: Experimental and computational studies," *Mater. Today Proc.*, vol. 18, pp. 1696–1708, 2019.

[23] E. A. M. Gad, E. M. S. Azzam, and S. A. Halim, "Theoretical approach for the performance of 4-mercapto-1-alkylpyridin-1-ium bromide as corrosion inhibitors using DFT," *Egypt. J. Pet.*, vol. 27, no. 4, pp. 695–699, 2018.

[24] J. Haque, V. Srivastava, C. Verma, and M. A. Quraishi, "Experimental and quantum chemical analysis of 2-amino-3-((4-((S)-2-amino-2-carboxyethyl)-1H-imidazol-2-yl)thio) propionic acid as new and green corrosion inhibitor for mild steel in 1 M hydrochloric acid solution," *J. Mol. Liq.*, vol. 225, no. February, pp. 848–855, 2017.

[25] M. Ouakki *et al.*, "Quantum chemical and experimental evaluation of the inhibitory action of two imidazole derivatives on mild steel corrosion in sulphuric acid medium," *Heliyon*, vol. 5, no. 11, 2019.

[26] S. K. Saha, M. Murmu, N. C. Murmu, I. B. Obot, and P. Banerjee, "Molecular level insights for the corrosion inhibition effectiveness of three amine derivatives on the carbon steel surface in the adverse medium: A combined density functional theory and molecular dynamics simulation study," *Surfaces and Interfaces*, vol. 10, no. 2017, pp. 65–73, 2018.

[27] D. K. Verma, "Density Functional Theory (DFT) as a Powerful Tool for Designing Corrosion Inhibitors in Aqueous Phase," in *Advanced Engineering Testing*, InTech, 2018.

[28] S. Hadisaputra, S. Hamdiani, and A. A. Purwoko, "Corrosion Inhibition of Steel in Acidic Medium by Eugenol Derivatives: Insight from Density Functional Calculation," *J. Appl. Chem. Sci.*, no. 2018, pp. 312–317, 2017.

[29] P. Udhayakala, "Density functional theory calculations on corrosion inhibitory action of five azlactones on mild steel," *Available online www.jocpr.com J. Chem. Pharm. Res.*, vol. 6, no. 7, pp. 117–127, 2014.

[30] P. Udhayakala and T. V Rajendiran, "A theoretical evaluation on benzothiazole derivatives as corrosion inhibitors on mild Steel," 2015.

[31] P. Mourya, P. Singh, A. K. Tewari, R. B. Rastogi, and M. M. Singh, "Relationship between structure and inhibition behaviour of quinolinium salts for mild steel corrosion: Experimental and theoretical approach," *Corros. Sci.*, vol. 95, pp. 71–87, 2015.

Chapter IV

CHAPTER IV

INVESTIGATION ON THE CORROSION INHIBITOR SYSTEM WITH WEIGHT LOSS MEASUREMENTS

4.1 Weight Loss Studies

The corrosion inhibition of mild steel in **1 M HCl** for the selected pharmaceutically active heterocyclic compounds, **BEX, CIL, MON, PYR** have been studied by weight loss method.

4.1.1 Effect of concentration on the inhibition efficiency of inhibitors

The inhibition efficiency with different concentrations **(5 – 60 mg/L)** of the inhibitors (BEX, CIL, MON & PYR) on mild steel in 1M HCl has been evaluated by weight loss measurements after different hours of immersion (3- 24 hours) and the results are given in the **table (1-4)**.

In this study, the inhibition efficiency for all the investigated pharmaceutically active heterocyclic drugs increases with increase in concentration [1]–[3]. Each inhibitor shows a maximum efficiency at the optimum concentration of 50 mg/L [4]–[6]. On further increase in the concentration to about 60mg/L there is no further increase in the efficiency of the studied drugs .The maximum efficiency of BEX, CIL, MON & PYR are 92%, 93%, 93% and 92% respectively in 1M HCl for either 6 or 12 hours of immersion time .

The results show that as the inhibitor concentration increases, the corrosion rate decreases. This implies that the studied inhibitors act through the adsorption on mild steel surface forming a barrier layer between the metal and corrosive medium [7], [8]. This may be due to the adsorption of heterocyclic system on to the mild steel surface through non-bonding electron pairs of hetero atoms as well as the π electrons of the aromatic rings [9], [10]. The high inhibitive performance of these compounds suggests a higher bonding ability of the inhibitor to the mild steel surface. In Monit the inhibition efficiency is due to the adsorption of the inhibitor on the mild steel surface through the nonbonding electron present on the hetero atom.

It is well known that in these compounds the lone pair of electrons of hetero atoms is responsible for much predominant adsorption on the mild steel surface. The protective property of these compounds is probably due to the interaction between π electrons of the heterocyclic system rings and hetero atoms with positively charged steel surface [11 - 12]

The corrosion inhibition can be attributed to the adsorption of inhibitor molecules at mild steel acid solution interface, blocking the active sites of acid attack and thereby protecting the metal from corrosion [13]–[15].

Table 4.1 Effect of concentration of BEX on the mild steel corrosion in 1M HCl at various time intervals

Conc mg/L	3 h		6 h		12 h		24h	
	CR mm/y	IE %	CR mm/y	IE %	CR mm/y	IE %	CR mm/y	IE %
Blank	68.59	-	72.22	-	96.78	-	275.54	-
5	55.27	19.41	56.86	21.27	74.24	23.29	220.57	19.95
10	51.25	25.27	44.89	37.85	41.81	56.80	186.94	32.15
15	40.84	40.46	29.42	59.27	34.74	64.10	145.51	47.19
20	32.66	52.39	23.50	67.46	25.81	73.33	83.32	69.76
25	30.50	55.53	21.24	70.60	25.74	73.41	76.18	72.35
30	25.52	62.80	21.57	70.13	19.94	79.40	60.63	78.00
35	21.28	68.98	20.83	71.16	18.75	80.63	60.40	78.08
40	19.64	71.37	15.21	78.94	12.87	86.70	39.05	85.83
45	19.12	72.13	12.90	82.13	8.18	91.54	31.47	88.58
50	16.81	75.49	11.75	83.73	7.44	92.31	29.98	89.12
55	10.19	80.12	10.86	84.96	7.22	92.54	23.06	91.63
60	9.97	**80.55**	10.52	**85.43**	7.07	**92.70**	22.39	**91.87**

Table 4.2 Effect of concentration of CIL on the mild steel corrosion in 1M HCl at various time intervals

Conc mg/L	3 h		6 h		12 h		24h	
	CR mm/y	IE %	CR mm/y	IE %	CR mm/y	IE %	CR mm/y	IE %
Blank	65.61		37.93		184.12	0	290.05	
5	46.34	29.37	28.00	26.18	89.34	51.47	168.79	41.81
10	44.71	31.86	24.73	34.80	71.12	61.37	161.28	44.40
15	39.87	39.23	24.43	35.59	64.79	64.81	146.40	49.53
20	38.76	40.93	20.27	46.57	46.20	74.91	136.73	52.86
25	37.12	43.42	19.12	49.61	43.96	76.12	129.59	55.32
30	36.23	44.78	18.63	50.88	39.50	78.55	120.51	58.45
35	33.25	49.32	13.39	64.71	31.32	82.99	111.73	61.48
40	29.68	54.76	11.94	68.53	27.90	84.85	95.74	66.99
45	24.18	63.15	10.75	71.67	22.39	87.84	85.70	70.45
50	22.24	66.10	8.18	78.43	14.88	91.92	79.67	72.53
55	21.20	67.69	6.29	83.43	13.54	92.65	75.28	74.04
60	19.71	**69.95**	5.84	**84.61**	11.90	**93.54**	67.84	**76.61**

Table 4.3 Effect of concentration of MON on the mild steel corrosion in 1M HCl at various time intervals

Conc mg/L	3 h		6 h		12 h		24h	
	CR mm/y	IE %	CR mm/y	IE %	CR mm/y	IE %	CR mm/y	IE %
Blank	65.61		37.93		184.12	0	364.44	
5	33.77	48.53	11.86	68.73	60.26	67.27	189.69	47.95
10	30.80	53.06	8.33	78.04	58.32	68.32	186.20	48.91
15	27.67	57.82	5.24	86.18	40.10	78.22	156.29	57.11
20	25.14	61.68	4.72	87.55	37.72	79.52	141.86	61.07
25	22.91	65.08	4.24	88.82	33.25	81.94	114.71	68.52
30	21.05	67.91	4.13	89.12	33.55	81.78	107.27	70.57
35	17.26	73.70	4.02	89.41	30.95	83.19	97.53	73.24
40	15.85	75.85	3.87	89.80	30.13	83.64	96.71	73.46
45	13.02	80.16	3.68	90.29	26.78	85.45	96.04	73.65
50	11.31	82.77	3.61	90.49	21.28	88.44	93.06	74.46
55	9.52	85.49	3.50	90.78	17.63	90.42	90.01	75.30
60	7.81	**88.10**	2.64	**93.04**	16.37	**91.11**	82.13	**77.46**

Table 4.4 Effect of concentration of PYR on the mild steel corrosion in 1M HCl at various time intervals

Conc mg/L	3 h		6 h		12 h		24h	
	CR mm/y	IE %	CR mm/y	IE %	CR mm/y	IE %	CR mm/y	IE %
Blank	68.59		54.52		184.12	0	364.44	
5	25.44	62.91	14.54	73.33	88.30	52.04	204.35	43.93
10	22.17	67.68	12.72	76.67	84.58	54.06	196.02	46.21
15	21.20	69.09	11.68	78.58	76.47	58.46	181.14	50.30
20	18.60	72.89	11.42	79.06	69.03	62.51	166.26	54.38
25	16.51	75.92	10.12	81.45	62.41	66.10	154.66	57.56
30	15.70	77.11	9.07	83.36	60.18	67.31	146.62	59.77
35	14.51	78.85	7.40	86.43	52.22	71.64	139.11	61.83
40	13.69	80.04	6.55	87.99	50.36	72.65	124.01	65.97
45	12.65	81.56	5.65	89.63	46.64	74.67	109.35	69.99
50	12.20	82.21	4.98	90.86	43.00	76.65	106.79	70.70
55	9.00	86.88	4.72	91.34	40.54	77.98	99.36	72.74
60	8.55	**87.53**	3.98	**92.70**	39.50	**78.55**	90.31	**75.22**

4.1.2. Effect of immersion time on the inhibition efficiency of the compounds

Metal samples were immersed in the acid medium for different duration of time also the inhibitor is used for repeated cleaning and remains in the acid solution for several hours. Thus the effect of exposure period on the inhibition efficiency for the heterocyclic derivatives at the maximum concentration of 60 mg/L.

The maximum IE of 92%, 93%, are noted in HCl for 6h of immersion time in the presence of compound BEX, CIL respectively, whereas an efficiency of 93% and 92% for

3h of immersion time in the presence of the compound MON & PYR respectively in HCl. Thereafter with increase in immersion time and the IE decreases to 91%, 76%, 77% and 75% in HCl for 24 h. As evident from the Figure 5, it is noted that as the immersion time increases from 3 h to 6 h, the inhibition efficiency increased in the acid medium for all the studied inhibitors. But there is a slight decrease in the inhibition efficiency after 6h in the case of MON & PYR and after 12h for BEX & CIL. The increase in corrosion rate at longer immersion time may be due to them desorption of the protective layer on the mild steel surface [16]–[18]

The optimum immersion period for all the studied heterocyclic derivatives at which maximum efficiency is reached is for 6h in HCl.

4.1.3. Effect of temperature on the inhibition efficiency of the compounds

Generally, the metallic corrosion in acidic media is accompanied by the evolution of H_2 gas and rises with temperature that usually accelerates the corrosion reactions, involving higher dissolution rate of the metal. The effect of temperature on the corrosion rate of mild steel in 1 M HCl solutions in the absence and presence of different concentrations (5 – 60 mg/L) of BEX, CIL, MON & PYR was studied at different temperatures (303, 313,323,333 & 343 K) by weight loss measurements for 1h immersion period.

The effect of temperature on the inhibited acid-metal reaction is very complex, because many changes occur on the metal surface such as rapid etching, desorption of inhibitor and the inhibitor itself may undergo decomposition and /or rearrangement. The inhibition efficiency is reduced to 87%, 72%, 74% & 62% at 343 K as compared to 97%, 93%, 96% & 94% at 323 K (60 mg/L) in 1M HCl for BEX, CIL, MON & PYR respectively.

Tables 5-8 shows that the inhibition efficiency decreases as the temperature increases and with decrease in concentrations of studied inhibitors [19]–[21]. Figure 6-9 shows the variation of percentage inhibition efficiency with temperature. It is clear from the figure that percentage inhibition efficiency increases with increase in concentration of pharmaceutically active heterocyclic compounds but with increase in temperature, the compounds shows a different trend in efficiency. Among most of the studied system, the inhibition efficiency increases initially with temperature from 303 K to 323 K. on further increase in temperature to 343K there is a decline in the inhibition efficiency.

The behavior of decrease in inhibition efficiency with increasing temperature is due to the adsorption and desorption of inhibitor molecules continuously occur at the metal surface and equilibrium exists between two processes at a particular temperature. With increase of temperature the equilibrium between adsorption and desorption processes is shifted to a higher desorption rate than adsorption until equilibrium is again establish at a different value of equilibrium constant. It explains the lower inhibition efficiency at higher temperature.

Table 4.5 Effect of temperature on the inhibition efficiency of BEX in 1M HCl from weight loss measurements

Concentration mg/L	303 K		313 K		323 K		333 K		343 K	
	CR mm/y	IE %	CR mm/y	IE %	CR mm/y	IE %	CR mm/y	IE %	CR mm/y	IE %
Blank	53.79		69.41		148.86		278.98		517.11	
5	42.85	20.33	33.25	52.09	43.74	70.61	57.13	79.52	166.05	67.89
10	39.28	26.97	31.02	55.31	19.42	86.96	29.01	89.60	125.20	75.79
15	34.82	35.27	27.45	60.45	18.75	87.41	25.44	90.88	112.26	78.29
20	32.58	39.42	27.23	60.77	18.30	87.71	24.55	91.20	53.79	89.60
25	30.35	43.57	26.56	61.74	17.85	88.01	18.52	93.36	47.09	90.89
30	27.67	48.55	23.43	66.24	13.84	90.70	14.95	94.64	43.97	91.50
35	26.56	50.62	22.09	68.17	11.38	92.35	14.73	94.72	39.95	92.27
40	24.33	54.77	20.53	70.42	8.03	94.60	14.28	94.88	32.14	93.79
45	23.66	56.02	19.86	71.38	7.36	95.05	13.61	95.12	32.58	93.70
50	22.76	57.68	19.19	72.35	6.47	95.65	12.27	95.60	16.52	96.81
55	20.98	61.00	18.75	72.99	5.36	96.40	11.83	95.76	16.29	96.85
60	20.73	**61.46**	18.30	**73.63**	4.46	**97.00**	11.38	**95.92**	16.07	**87.17**

Table 4.6 Effect of temperature on the inhibition efficiency of CIL in 1M HCl from weight loss measurements

Conc mg/L	303 K		313 K		323 K		333 K		343 K	
	CR mm/y	IE %	CR mm/y	IE %	CR mm/y	IE %	CR mm/y	IE %	CR mm/y	IE %
Blank	19.19		57.13		183.45		399.49		519.79	
5	14.28	25.58	26.78	53.12	65.39	64.36	229.43	42.57	371.82	28.47
10	13.39	30.23	24.55	57.03	65.17	64.48	206.22	48.38	364.01	29.97
15	12.72	33.72	21.20	62.89	57.36	68.73	171.85	56.98	354.63	31.77
20	12.05	37.21	19.86	65.23	48.21	73.72	135.47	66.09	335.22	35.51
25	11.61	39.53	17.18	69.92	45.08	75.43	124.76	68.77	319.37	38.56
30	11.16	41.86	16.29	71.48	27.67	84.91	76.77	80.78	295.04	43.24
35	10.71	44.19	14.73	74.22	25.89	85.89	71.19	82.18	261.34	49.72
40	10.27	46.51	13.61	76.17	22.54	87.71	66.51	83.35	242.60	53.33
45	9.37	51.16	11.61	79.69	22.09	87.96	65.84	83.52	219.83	57.71
50	8.70	54.65	10.71	81.25	17.18	90.63	48.65	87.82	191.49	63.16
55	8.26	56.98	9.82	82.81	14.06	92.34	45.31	88.66	174.30	66.47
60	7.81	**59.30**	9.37	**83.59**	11.61	**93.67**	64.95	**83.74**	142.61	**72.56**

Table 4.7 Effect of temperature on the inhibition efficiency of MON in 1M HCl from weight loss measurements

Conc mg/L	303 K		313 K		323 K		333 K		343 K	
	CR mm/y	IE %	CR mm/y	IE %	CR mm/y	IE %	CR mm/y	IE %	CR mm/y	IE %
Blank	59.37		101.77		183.23		366.24		k324.50	
5	44.19	25.56	43.07	57.68	59.37	67.60	149.53	59.17	176.09	45.74
10	38.39	35.34	33.25	67.32	30.58	83.31	141.72	61.30	163.81	49.52
15	31.47	46.99	27.45	73.03	19.42	89.40	131.68	64.05	157.57	51.44
20	28.79	51.50	22.32	78.07	19.42	89.40	119.85	67.28	143.50	55.78
25	24.77	58.27	19.64	80.70	15.40	91.60	107.80	70.57	136.14	58.05
30	24.33	59.02	17.41	82.89	14.51	92.08	106.01	71.05	119.85	63.07
35	22.76	61.65	16.29	83.99	13.17	92.81	96.64	73.61	104.00	67.95
40	21.65	63.53	14.51	85.75	10.49	94.28	81.01	77.88	96.64	70.22
45	20.31	65.79	12.27	87.94	8.93	95.13	75.43	79.40	89.94	72.28
50	19.86	66.54	11.16	89.04	6.70	96.35	66.51	81.84	84.59	73.93
55	18.30	69.17	9.82	90.35	6.47	96.47	64.28	82.45	82.58	74.55
60	17.63	**70.30**	9.60	**90.57**	6.25	**96.59**	59.37	**83.79**	81.68	**74.83**

Table 4.8 Effect of temperature on the inhibition efficiency of PYR in 1M HCl from weight loss measurements

Conc mg/L	303 K		313 K		323 K		333 K		343 K	
	CR mm/y	IE %	CR mm/y	IE %	CR mm/y	IE %	CR mm/y	IE %	CR mm/y	IE %
Blank	59.37		78.78		168.50		256.66		519.79	
5	43.97	25.94	37.05	52.97	129.44	23.18	213.14	16.96	473.59	8.89
10	41.73	29.70	19.42	75.35	121.19	28.08	199.75	22.17	473.36	8.93
15	32.14	45.86	18.75	76.20	98.42	41.59	165.15	35.65	407.53	21.60
20	27.90	53.01	18.30	76.77	93.74	44.37	152.21	40.70	264.69	49.08
25	25.67	56.77	17.85	77.34	79.01	53.11	150.20	41.48	249.74	51.95
30	20.53	65.41	13.84	82.44	68.52	59.34	138.59	46.00	242.60	53.33
35	18.08	69.55	11.38	85.55	68.29	59.47	111.81	56.43	226.30	56.46
40	15.85	73.31	8.03	89.80	53.12	68.48	93.07	63.74	216.93	58.27
45	13.61	77.07	7.36	90.65	47.76	71.66	76.55	70.17	212.24	59.17
50	11.83	80.08	6.47	91.78	44.86	73.38	72.76	71.65	193.72	62.73
55	11.61	80.45	5.36	93.20	42.40	74.83	70.08	72.70	186.36	64.15
60	11.16	**81.20**	4.46	**94.33**	24.10	**85.70**	57.13	**77.74**	178.77	**65.61**

4.1.4. Comparison of the corrosion inhibition behavior of the studied inhibitors

The additives are found to be more effective at 323K (higher temperature) than 333K and 343 K, signifying that the compounds are strongly adsorbed on the mild steel coupons.

The order of decreasing the inhibition efficiency of the investigated compounds in the corrosive solutions was as follows:

BEX > CIL > MON > PYR

It is evident that the inhibition efficiency of BEX, CIL, MON & PYR depends mainly on the molecular size of the compounds, the charge density on the adsorption sites, electron clouds, mode of interaction with the metal surface and formation of metallic complexes.

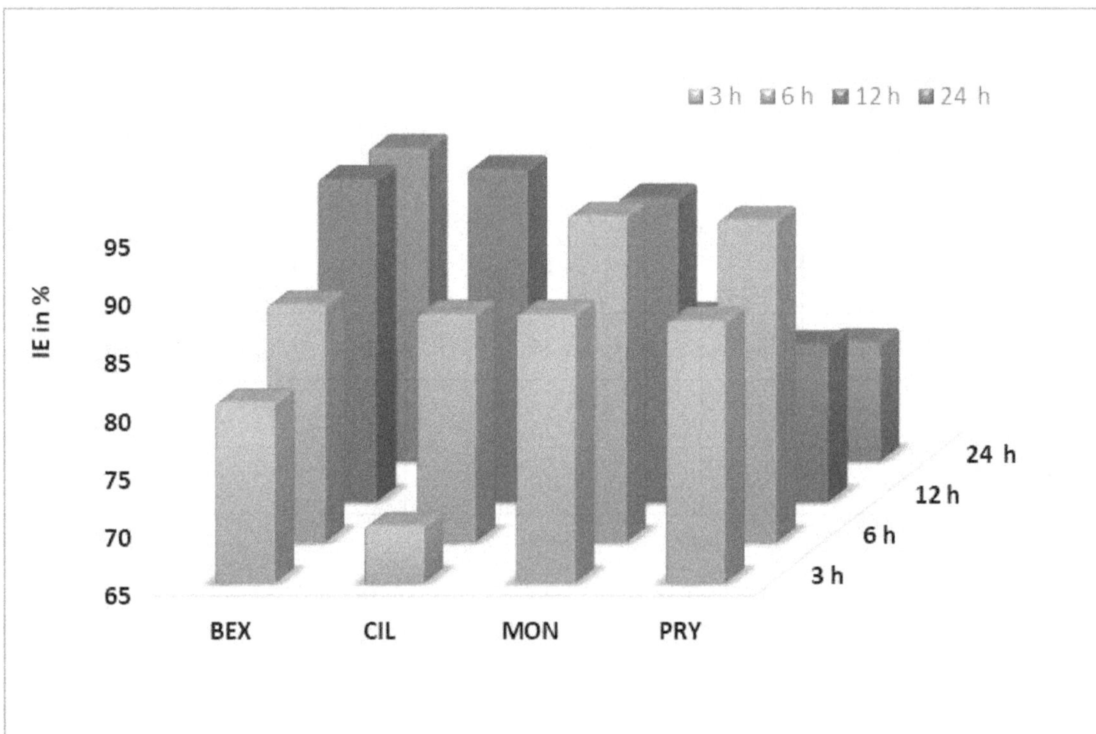

Figure 4.1 Comparison of the corrosion inhibition behavior of the studied inhibitors in 1M HCl

4.1.5. Adsorption Isotherm

Adsorption occurs, metallic surface of the valence electrons undergo chemical bonding and electrostatic attraction and /or heteroatoms of the compound [22] . Which consist a protective layer formation and water molecules removals. Complex of these mechanisms depends on the functional group,

Organic substance adsorbed to the metal solution interface, this is written as

$$\textbf{Org (ads) + nH2O} \rightarrow \textbf{org (ads) + nH2O} \qquad \textbf{1}$$

Where n is the number of H2O molecules removed from the surface of the metal. Value of 'n' will rest on the organic molecule area with respect to that of the H2O molecule.

Reabsorbed of the water molecule to the metal surface interaction of the aqueous solution are elaborated in the adsorption process.

Main function of the corrosion inhibitors is, intrusive the cathodic or anodic or both the reactions. Corrosion inhibition process is generally obtained by the adsorption of the metal - solution interface of the neutral polar molecules or ion and solution and metal electrode of the potential difference is due to the interface of the non-uniform distribution of the electric charges. Electrical double layer of the ions or neutral molecule interaction, changed its structure and properties. In adsorption process reabsorbed of the water molecule at the metal surface an interaction through the aqueous solution.

In order to get a better understanding of the corrosion process on the metal surface, adsorption characteristics are also studied. This process is closely associated to the adsorption of the inhibitor molecules and its chemical structure is related to adsorption. Adsorption isotherms, act as an important role for the defining the mechanism of electrochemical reactions

The adsorption method is subjective by the environment and surface charge of the metal, inhibitors chemical structure and the type of destructive electrolyte. From the present study, the gravimetric data were fitted to the most often used adsorption isotherms, namely: **Bockris Swinkles, Hillde Boer, Dhar-Flory Huggins, Pausons, and Langmuir, Freundlich, Temkin, Frumkin,** the best fit to correlation coefficients (R^2) were used to regulate [23]–[25]. General form of all of these isotherms.

$$f(o, x)\exp(-2a\theta) = Kc \qquad 2$$

where f(o, x) is the configurational factor which depends upon the physical model and fundamental the derivative of the isotherm, 'a' is the molecular interaction parameter ,θ is the surface coverage, k is the equilibrium constant of adsorption process. c is the inhibitor coverage, x is the size ration . By far, the Langmuir's and Temkin adsorption isotherm are best fitted to its experimental data at 303K, 313K, 323K, 333K, and 343 K. The degree of surface coverage (θ) was calculated, in the presence of inhibitor at different concentrations (5mg/L to 60mg/L) at 303K after immersion for 3h, 6h, 12h, and 24h in 1M HCl was calculated from weight loss data.

i) Langmuir adsorption isotherm:

A relationship between surface coverage of the inhibitor and concentration in an electrolyte characterize by the following equation

$$\frac{C}{\theta} = \frac{1}{K_{ads}} + C \qquad 3$$

where c is the inhibitor concentration in ppm, K_{ads} is the equilibrium constant for adsorption–desorption process, θ is the surface coverage. Thus for the plot of C/θ versus C, straight line was obtained in the case of Langmuir adsorption isotherm.

Figure (4.2-4.5) shows the plot of C/θ Vs C and the expected linear connection is obtained with R^2 values close to unity for all the studied inhibitors, BEX, CIL, MON and PYR in 1M HCl. The R^2 values were very close to unity indicating strong adherence of adsorbed film of the inhibitor molecules to the metal surface which fits to the Langmuir adsorption isotherm.

Langmuir isotherm proposes monolayer layer adsorption at specific reaction sites on the steel's surfaces [26 - 27] The adsorptions are identical, equivalent and confirms no cross interaction between the adsorbed molecules at the metal/solution interface.

ii) Temkin adsorption isotherm:

The Temkin adsorption isotherm was also fitted to describe the experimental results for BEX, CIL, MON and PYR adsorption on the mild steel surface at 303K as characterized by the equation

$$\theta = log\ (Corrosion\ Rate) + K_{ads} \qquad 4$$

According the above equation, the plot of Θ as a function of log C gives a straight line with intercept equals to K_{ads} [28 - 30]. The Temkin plot for all the studied inhibitors, BEX, CIL, MON and PYR in 1M HCl was depicted in Figure 4.6-4.9 which shows a good correlation for the experimental results.

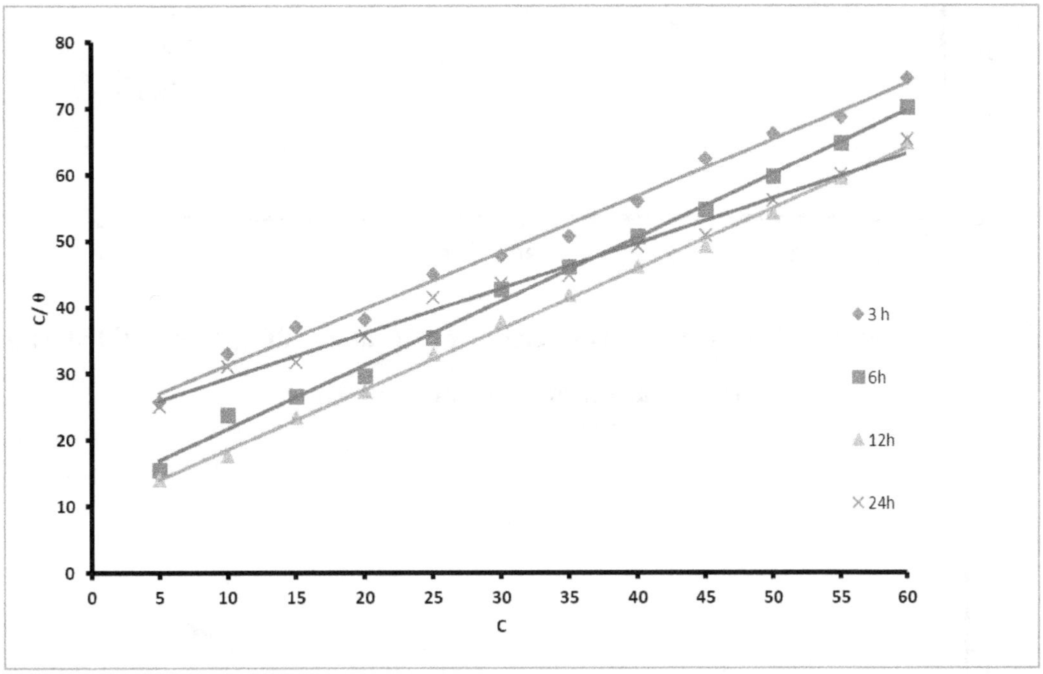

Figure 4.2 Langmuir adsorption isotherm for BEX at 5- 60mg/L on mild steel in 1M HCl at different immersion time.

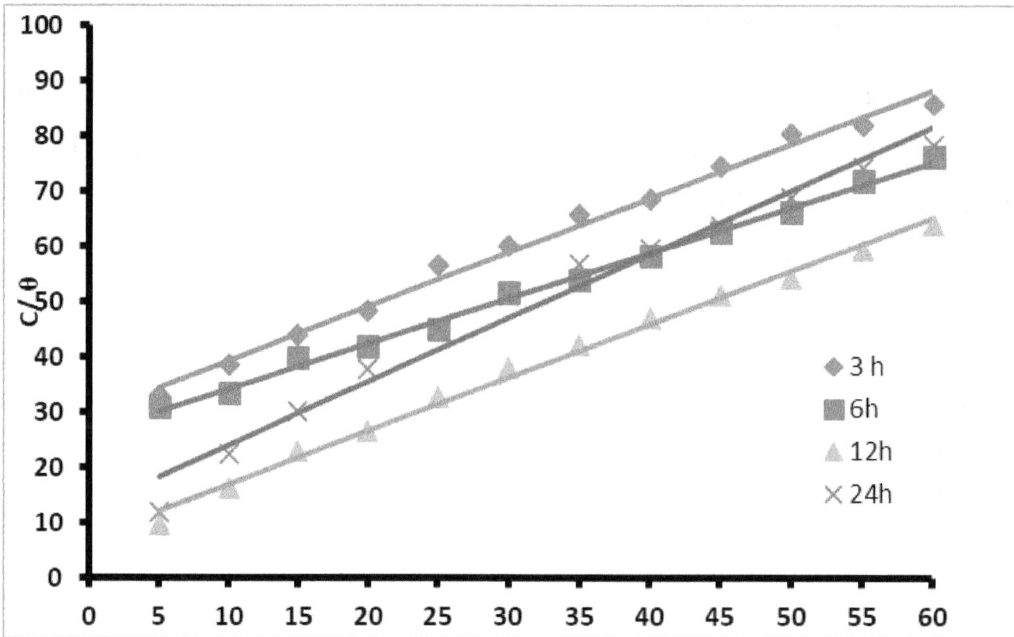

Figure 4.3 Langmuir adsorption isotherm for CIL at 5- 60mg/L on mild steel in 1M HCl at different immersion time.

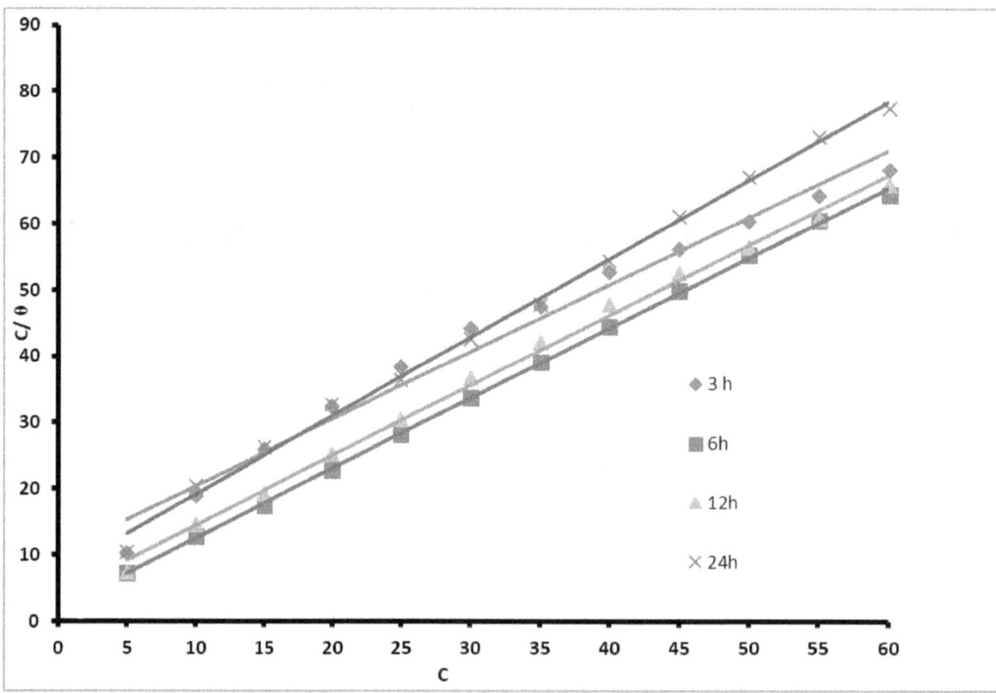

Figure 4.4 Langmuir adsorption isotherm for MON at 5- 60mg/L on mild steel in 1M HCl at different immersion time

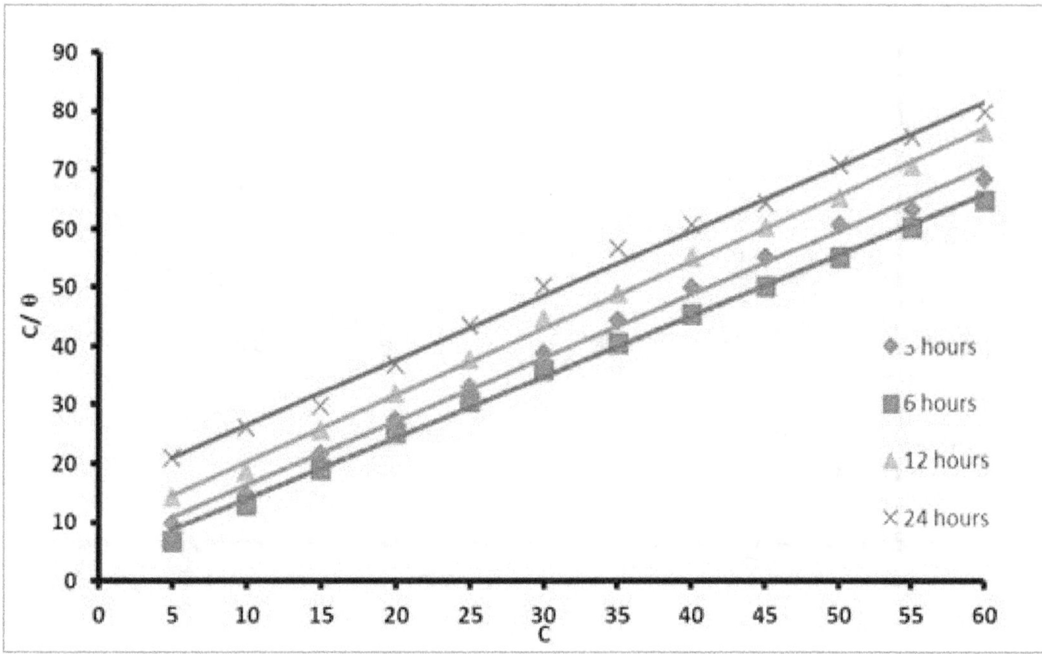

Figure 4.5 Langmuir adsorption isotherm for PYR at 5- 60mg/L on mild steel in 1M HCl at different immersion time.

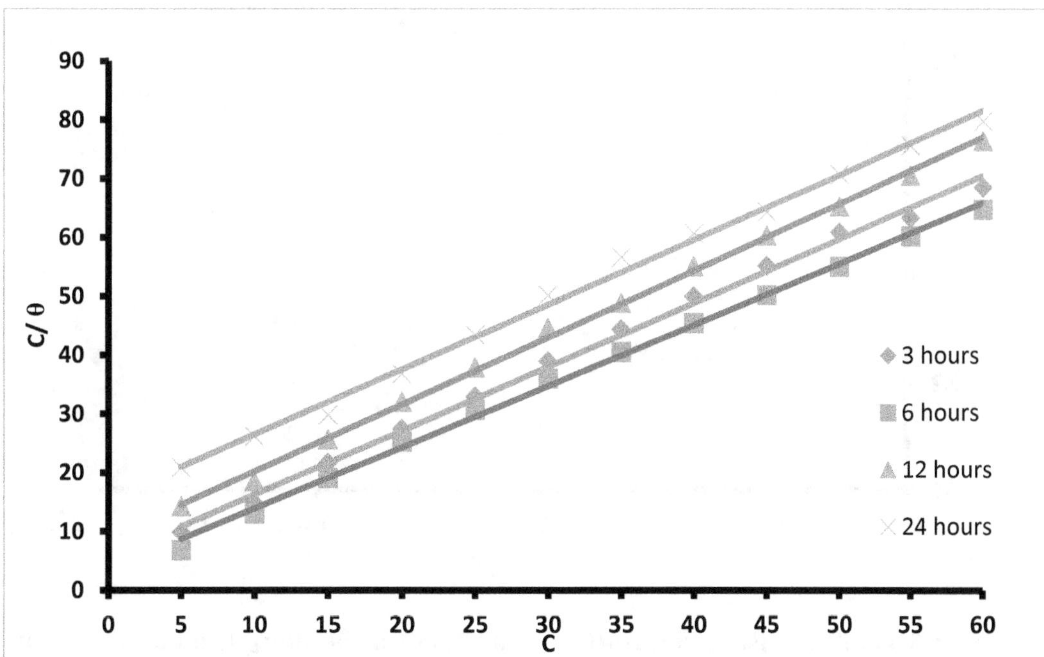

Figure 4.6 Temkin adsorption isotherm for BEX at 5- 60mg/L on mild steel in 1M HCl at different immersion time.

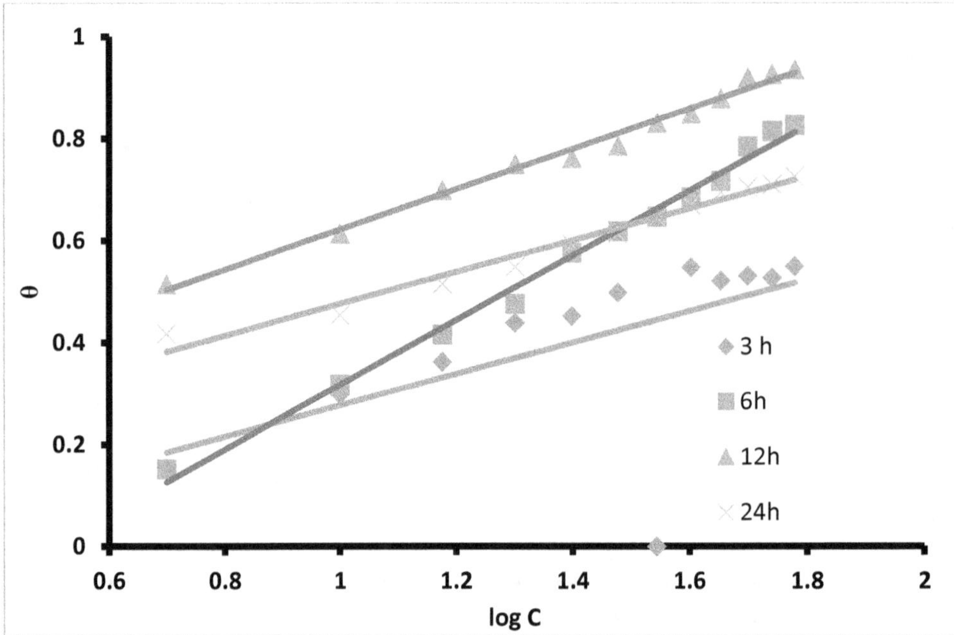

Figure 4.7 Temkin adsorption isotherm for CIL at 5- 60mg/L on mild steel in 1M HCl at different immersion time.

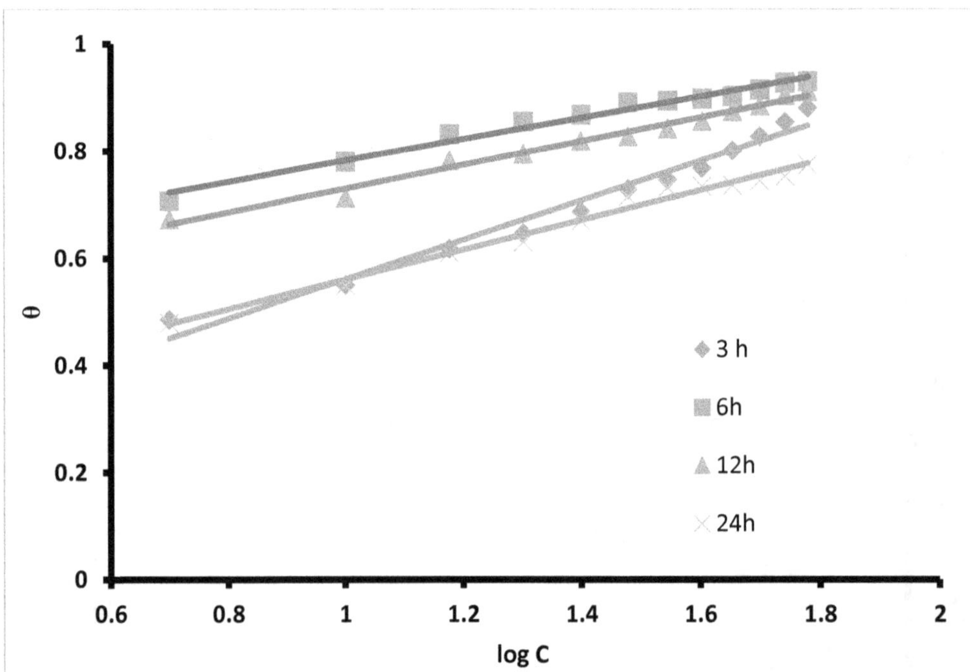

Figure 4.8 Temkin adsorption isotherm for MON at 5- 60mg/L on mild steel in 1M HCl at different immersion time.

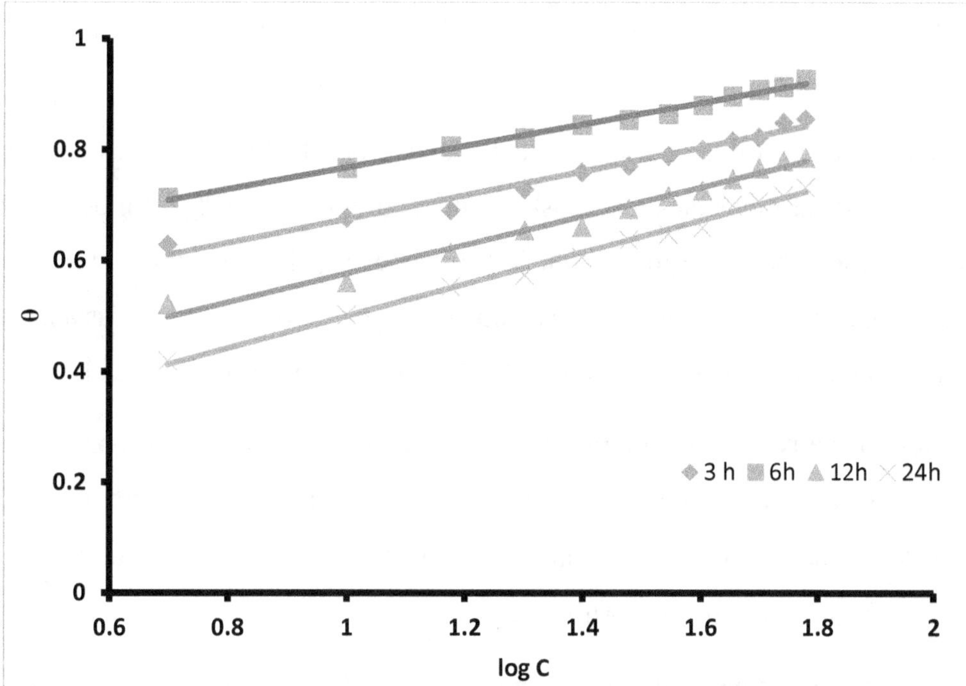

Figure 4.9 Temkin adsorption isotherm for PYR at 5- 60mg/L on mild steel in 1M HCl at different immersion time.

iii) Adsorption equilibrium Constant (K_{ads})

Values of adsorption equilibrium constant are obtained from the intercept of Langmuir adsorption isotherm [31]. The adsorption constant have high values, improved inhibition efficiency and mild steel surface are strongly adsorbed**.** This due to the molecular structure of the inhibitors, have the presence of π-electrons. Langmuir adsorption isotherm is connected to free energy of adsorption

Adsorbed layer is formed on the metal surface, this shows that the values of Kads. It also signifies that the strength of the adsorbent and adsorbate. Replaced of water molecules by the adsorbed species through the interaction, this is shows that small deviances in the arc.

$$\Delta G_{ads} = -\boldsymbol{RTln}(\mathbf{55.5}\boldsymbol{K})$$

where $\Delta G_{ads}$the free energy of adsorption, T is the absolute temperature (K), R is the gas constant, the values of 55.5 is the concentration of water in solution in mol L^{-1}. The negative values of $\boldsymbol{\Delta G_{ads}}$ suggest that, the adsorption process is spontaneous and exothermic environment of the process of corrosion. The calculated values of K_{ads} and ΔG_{ads} are shown in **table 4.9.**

Generally, the value of $\boldsymbol{\Delta G_{ads}}$ is less negative or 20 kJ/mol. At the charged metal surface and charged molecules, electrostatic interaction occurs. (Physical adsorption). It has been suggested that adsorption of the negatively charged of the inhibitor is enabled to the positive charge of the metal. But the positive charge of the inhibitor act as a negatively charge intermediates such as acid anions adsorbed on the surface of the metal.

More negative or 40kJ/mol involves a co-ordinate covalent bond (chemical adsorption) is formed between metal surface and transfer of the inhibitor molecule. Standard free energy of adsorption ($\Delta G_{ads)}$ calculated values is less than 40 kJ/mol it obeys that the adsorption is physical adsorption.

From the experiment, the ΔG_{ads} value of BEX, CIL, MON and PYR were is less than -40kJ/mol. The investigation is showed that the adsorption all the inhibitors on the surface of the mild steel followed by physisorption mechanism.

Table 4.9 Thermodynamic parameters for the adsorption inhibitors of BEX, CIL, MON and PYR in different concentrations of mild steel in 1M HCl

Immersion time (hours)	Langumir Adsorption Isotherm				Temkin Adsorption Isotherm		
	R^2	Intercept	K_{ads}	ΔG°_{ads}	R^2	K_{ads}	ΔG°_{ads}
BEX							
3	0.992	22.771	0.0439	-2.24	0.991	0.2755	-6.87
6	0.995	12.093	0.0826	-3.84	0.986	0.1869	-10.11
12	0.998	9.4023	0.1063	-4.47	0.990	0.0418	-2.11
24	0.987	22.49	0.0444	-2.27	0.9914	0.2703	-6.82
CIL							
3	0.9918	29.755	0.0336	-1.57	0.963	0.0093	-10.12
6	0.995	26.036	0.0384	-1.91	0.991	0.3177	-10.11
12	0.993	7.5353	0.1327	-5.03	0.989	0.2265	-6.38
24	0.976	12.697	0.0787	-3.71	0.974	0.1635	-5.56
MON							
3	0.981	10.291	0.0971	-4.24	0.972	0.1916	-5.96
6	0.9995	1.9277	0.5187	-8.47	0.977	0.5839	-8.76
12	0.9966	3.8552	0.2593	-6.71	0.9846	0.5077	-8.41
24	0.9966	7.2364	0.1381	-5.13	0.982	0.2823	-6.93
PYR							
3	0.9962	5.3803	0.1858	-5.88	0.974	0.4605	-8.16
6	0.9972	3.4724	0.2879	-6.98	0.993	0.5730	-8.72
12	0.9982	8.8075	0.1135	-4.64	0.981	0.3165	-7.22
24	0.9954	15.461	0.0646	-3.21	0.992	0.2105	6.19

References

[1] A. Khadiri *et al.*, "Gravimetric, electrochemical and quantum chemical studies of some pyridazine derivatives as corrosion inhibitors for mild steel in 1 M HCl solution," *J. Taiwan Inst. Chem. Eng.*, vol. 58, pp. 552–564, 2016.

[2] S. Al-Baghdadi, T. S. Gaaz, A. Al-Adili, A. A. Al-Amiery, and M. S. Takriff, "Experimental studies on corrosion inhibition performance of acetylthiophene thiosemicarbazone for mild steel in HCl complemented with DFT investigation," *Int. J. Low-Carbon Technol.*, vol. 16, no. 1, pp. 181–188, 2021.

[3] K. R. Ansari, S. Ramkumar, D. Nalini, and M. A. Quraishi, "Studies on adsorption and corrosion inhibitive properties of quinoline derivatives on N80 steel in 15% hydrochloric acid," *Cogent Chem.*, vol. 2, no. 1, p. 1145032, 2016.

[4] I. Ahamad, R. Prasad, and M. A. Quraishi, "Thermodynamic, electrochemical and quantum chemical investigation of some Schiff bases as corrosion inhibitors for mild steel in hydrochloric acid solutions," *Corros. Sci.*, vol. 52, no. 3, pp. 933–942, 2010.

[5] A. Singh *et al.*, "Effect of electron donating functional groups on corrosion inhibition of mild steel in hydrochloric acid: Experimental and quantum chemical study," *J. Taiwan Inst. Chem. Eng.*, vol. 82, pp. 233–251, 2018.

[6] F. E. Abeng, V. C. Anadebe, P. Y. Nkom, K. J. Uwakwe, and E. G. Kamalu, "Experimental and theoretical study on the corrosion inhibitor potential of quinazoline derivative for mild steel in hydrochloric acid solution," *J. Electrochem. Sci. Eng.*, vol. 11, no. 1, pp. 11–26, 2021.

[7] N. Parhizkar, T. Shahrabi, and B. Ramezanzadeh, "A new approach for enhancement of the corrosion protection properties and interfacial adhesion bonds between the epoxy coating and steel substrate through surface treatment by covalently modified amino functionalized graphene oxide film," *Corros. Sci.*, vol. 123, no. April, pp. 55–75, 2017.

[8] M. Honarvar Nazari, M. S. Shihab, E. A. Havens, and X. Shi, "Mechanism of corrosion protection in chloride solution by an apple-based green inhibitor: experimental and theoretical studies," *J. Infrastruct. Preserv. Resil.*, vol. 1, no. 1, 2020.

[9] R. S. Abdel Hameed, A. H. Al-Bagawi, H. A. Shehata, A. H. Shamroukh, and M. Abdallah, "Corrosion Inhibition and Adsorption Properties of Some Heterocyclic Derivatives on C-Steel Surface in HCl," *J. Bio- Tribo-Corrosion*, vol. 6, no. 2, 2020.

[10] D. T. Oyekunle, O. Agboola, and A. O. Ayeni, "Corrosion Inhibitors as Building Evidence for Mild Steel: A Review," in *Journal of Physics: Conference Series*, 2019, vol. 1378, no. 3.

[11] P. Mourya, P. Singh, A. K. Tewari, R. B. Rastogi, and M. M. Singh, "Relationship between structure and inhibition behaviour of quinolinium salts for mild steel corrosion: Experimental and theoretical approach," *Corros. Sci.*, vol. 95, pp. 71–87, 2015.

[12] T. K. Chaitra, K. N. Mohana, and H. C. Tandon, "Evaluation of newly synthesized hydrazones as mild steel corrosion inhibitors by adsorption, electrochemical, quantum chemical and morphological studies," *Arab J. Basic Appl. Sci.*, vol. 25, no. 2, pp. 45–55, 2018.

[13] E. Kowsari *et al.*, "In situ synthesis, electrochemical and quantum chemical analysis of an amino acid-derived ionic liquid inhibitor for corrosion protection of mild steel in 1M HCl solution," *Corros. Sci.*, vol. 112, pp. 73–85, 2016.

[14] C. chuan Li *et al.*, "Adsorption and corrosion inhibition of phytic acid calcium on the copper surface in 3wt% NaCl solution," *Corros. Sci.*, vol. 83, pp. 147–154, 2014.

[15] C. Verma, D. S. Chauhan, and M. A. Quraishi, "Drugs as environmentally benign corrosion inhibitors for ferrous and nonferrous materials in acid environment: An overview," *J. Mater. Environ. Sci.*, vol. 8, no. 11, pp. 4040–4051, 2017.

[16] S. Al-Baghdadi, T. S. Gaaz, A. Al-Adili, A. A. Al-Amiery, and M. S. Takriff, "Experimental studies on corrosion inhibition performance of acetylthiophene thiosemicarbazone for mild steel in HCl complemented with DFT investigation," *Int. J. Low-Carbon Technol.*, vol. 16, no. 1, pp. 181–188, 2021.

[17] N. Raghavendra, "Expired abacavir sulfate drug as non-toxic corrosion inhibitor for mild steel (MS) in 3 M hydrochloric acid system," *Gazi Univ. J. Sci.*, vol. 32, no. 4, pp. 1113–1121, 2019.

[18] A. Pradityana, Sulistijono, A. Shahab, L. Noerochim, and D. Susanti, "Inhibition of Corrosion of Carbon Steel in 3.5% NaCl Solution by Myrmecodia Pendans Extract," *Int. J. Corros.*, vol. 2016, 2016.

[19] R. Mohan and A. Joseph, "Corrosion protection of mild steel in hydrochloric acid up to 313 K using propyl benzimidazole: Electroanalytical, adsorption and quantum chemical studies," *Egypt. J. Pet.*, vol. 27, no. 1, pp. 11–20, 2018.

[20] A. S. Fouda, M. A. El Morsi, and T. El Mogy, "Studies on the inhibition of carbon steel corrosion in hydrochloric acid solution by expired carvedilol drug," *Green Chem. Lett. Rev.*, vol. 10, no. 4, pp. 336–345, 2017.

[21] S. Ramkumar and D. Nalini, "Correlation between inhibition efficiency and chemical structure of new indolo imidazoline on the corrosion of mild steel in molar HCl with DFT evidences," *Orient. J. Chem.*, vol. 31, no. 2, pp. 1057–1064, 2015.

[22] S. Ramkumar and D. Nalini, "Enhanced corrosion resistance of mild steel in sulphuric acid medium by imidazole derivative: Experimental and computational studies," *Mater. Today Proc.*, vol. 18, pp. 1696–1708, 2019.

[23] M. Chafiq *et al.*, "Naproxen-based hydrazones as effective corrosion inhibitors for mild steel in 1.0 M HCl," *Coatings*, vol. 10, no. 7, 2020.

[24] K. R. Ansari, M. A. Quraishi, and A. Singh, "Isatin derivatives as a non-toxic corrosion inhibitor for mild steel in 20% H2SO4," *Corros. Sci.*, vol. 95, no. December 2017, pp. 62–70, 2015.

[25] A. Nahlé *et al.*, "Novel triazole derivatives as ecological corrosion inhibitors for mild steel in 1.0 M HCl: experimental & theoretical approach," *RSC Adv.*, vol. 11, no. 7, pp. 4147–4162, 2021.

[26] X. Luo, X. Pan, S. Yuan, S. Du, C. Zhang, and Y. Liu, "Corrosion inhibition of mild steel in simulated seawater solution by a green eco-friendly mixture of glucomannan (GL) and bisquaternary ammonium salt (BQAS)," *Corros. Sci.*, vol. 125, no. June, pp. 139–151, 2017.

[27] P. Arockiasamy, X. Q. R. Sheela, G. Thenmozhi, M. Franco, J. W. Sahayaraj, and R. J. Santhi, "Evaluation of Corrosion Inhibition of Mild Steel in 1 M Hydrochloric Acid Solution by Mollugo cerviana," *Int. J. Corros.*, vol. 2014, 2014.

[28] D. C. Sinica, "Pelagia Research Library," vol. 6, no. 5, pp. 96–103, 2015.

[29] M. Ouknin, A. Romane, J. P. Ponthiaux, J. Costa, and L. Majidi, "Evaluation of corrosion inhibition and adsorption behavior of Thymus zygis subsp. gracilis volatile compounds on mild steel surface in 1 m HCl," *Corros. Rev.*, vol. 38, no. 2, pp. 137–149, 2020.

[30] M. Ouknin, A. Romane, J. P. Ponthiaux, J. Costa, and L. Majidi, "Evaluation of corrosion inhibition and adsorption behavior of Thymus zygis subsp. gracilis volatile compounds on mild steel surface in 1 m HCl," *Corros. Rev.*, vol. 38, no. 2, pp. 137–149, 2020.

[31] A. Fitoz, H. Nazır, M. Özgür (nee Yakut), E. Emregül, and K. C. Emregül, "An experimental and theoretical approach towards understanding the inhibitive behavior of a nitrile substituted coumarin compound as an effective acidic media inhibitor," *Corros. Sci.*, vol. 133, no. April , 2017, pp. 451–464, 2018.

Chapter V

CHAPTER V

EVALUATION OF CORROSION POTENTIAL OF THE STUDIED DRUGS BY ELECTROCHEMICAL METHODS

5.1. Electrochemical impedance spectroscopic measurements:

Electrochemical impedance measurements were carried out over the frequency range from 100 Kz to 0.01Hz at open circuit potential. Inhibition efficiency can be calculated from Nyquist plot as follows:

$$\mathbf{IE\ (\%)} = \frac{R^{i}_{ct} - R_{ct}}{R^{i}_{ct}} \qquad \mathbf{1}$$

where, $R^{i}ct$ and Rct are the charge transfer resistance of mild steel with and without inhibitor molecules respectively. Inhibition efficiencies and other calculated impedance parameters are presented in **Table 5.1.1 - 5.1.4**.

EIS technique was also used to investigate the corrosion behavior of mild steel in all the above conditions. Figures 5.1.1-5.1.4 show the complex plane impedance plots (Nyquist plots) for mild steel in 1M HCl with and without various concentrations of BEX, CIL, MON and PYR at 30°C. According to the electrochemical theory, Rct is inversely proportional to the corrosion current density (i.e) corrosion rate [1]–[4]. All experimental results obtained from EIS measurements may be interpreted as follows:

5.1.1. BEX

The Nyquist plots for **BEX** in 1M HCl is depicted in **Figure 5.1.1** and the impedance behavior is concise as given in **Table 5.1.1** is discussed as follows:

- R_{ct} increase from 19 to 201 Ωcm^2 in 1M HCl respectively.
- C_{dl} decreases from **12.9** to 11.8 $\mu F/cm^2$ at 120mg/L concentration in HCl.
- Increase in R_{ct} and decrease in C_{dl} increases the IE calculated from **57.78%** to **90.55%** in HCl

Table 5.1.1: Electrochemical impedance parameters for corrosion of mild steel at various concentrations of compound BEX in 1M HCl

Conc mg/L	Rs Ω cm²	Yo µF/cm²	n	Rct Ω cm²	IE %	fmax µHz	Cdl µF/cm²	θ	Sⁿ	Icor (from Rct) µA/cm²	IE %
1M HCl											
Blank	6.81	8076	0.25	19	-	6.52	12.9	-	0.24	1238806	-
5	9.23	11976	-0.41	45	57.78	1.86	19.1	-0.4829	7.47	290945.9	76.51
15	16.6	10984	-0.45	68	72.06	1.34	17.5	-0.3600	6.49	324130.8	73.84
30	16.1	10714	-0.50	107	82.24	0.87	17.1	-0.3266	1.02	205989.7	83.37
45	19.2	10469	-0.51	114	83.33	0.84	16.7	-0.2963	4.74	97380.62	92.14
90	20.3	8679	-0.54	135	85.93	0.85	13.8	-0.0747	13.5	82536.82	93.34
120	22.5	7379	-0.60	201	**90.55**	0.67	11.8	0.0863	14.29	65367.09	**94.72**

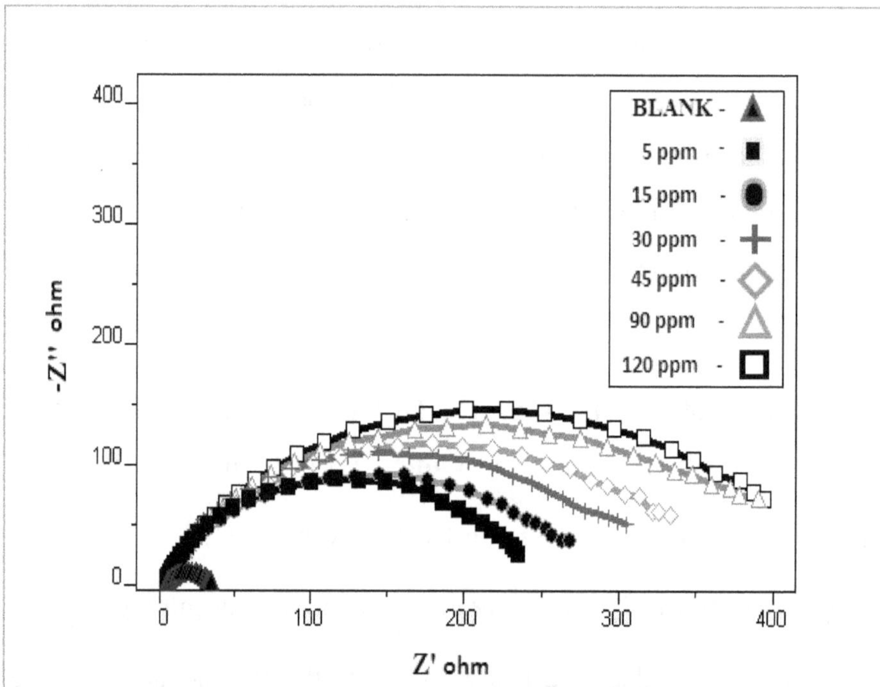

Figure 5.1.1 Nyquist curves recorded for mild steel electrode in 1M HCl Solution containing different concentrations of BEX

5.1.2. CIL

With the addition of compound **CIL** to the acid medium, the impedance behavior **Figure 5.1.2 & Table 5.1.2** is briefed as follows:

- Analysis of the data reveals that, the R_{ct} values increase from **19** Ohmcm^2 for the blank acid to **201** Ohmcm^2 thereby increasing the IE from **58.04%** to **90.61%** in HCl.
- C_{dl} decreases from **12.9** to **1.02** μF/cm^2 for blank HCl at 120mg/L.
- The decrease in C_{dl} values increases the surface coverage (θ) from **-1.4728** to **0.0700** in HCl solution
- The decrease in C_{dl} and increase in R_{ct} may be due to gradual replacement of water molecules by the adsorbed inhibitor [5]–[7].

Table 5.1.2: Electrochemical impedance parameters for corrosion of mild steel at various concentrations of compound CIL in 1M HCl

Conc mg/L	R_S Ω cm^2	Y_0 μF/cm^2	n	R_{ct} Ω cm^2	IE %	fmax μHz	C_{dl} μF/cm^2	θ	S^n	Icor (from Rct) μA/cm^2	IE %
1M HCl											
Blank	6.81	8076	0.25	19	-	6.56	12.9	-	0.24	1238806	-
5	9.23	12943	-0.40	45	58.04	1.11	3.18	-1.4728	1.43	482289	61.07
15	16.6	7511	-0.47	68	72.42	0.75	3.12	-1.4253	2.14	348138	71.90
30	16.1	19970	-0.47	107	82.40	0.49	3.05	-1.3717	3.27	215602	82.60
45	19.2	19154	-0.48	114	83.39	0.51	2.75	-1.1407	3.13	192826	84.44
90	20.3	19587	-0.50	135	86.03	0.57	2.06	-0.6026	2.78	164992	86.68
120	22.5	17289	-0.54	201	**90.61**	0.66	1.02	0.0700	2.4	113159	**90.87**

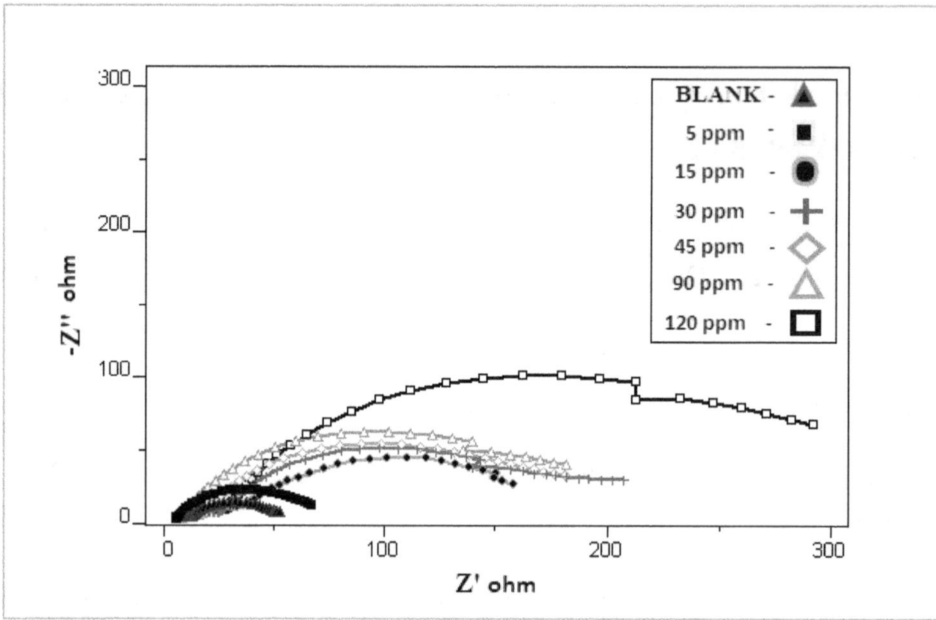

Figure 5.1.2. Nyquist curves recorded for mild steel electrode in 1M HCl Solution containing different concentrations of CIL

5.1.3. MON

The impedance behavior **Figure 5.1.3 & Table 5.1.3**) with the addition of compound **MON** to the acid medium is summarized as follows:

- The diameter of Nyquist plot increase on increasing the concentration where the R_{ct} value increase from **19** and **201** Ohmcm^2 for the blank acid to in HCl respectively [8]–[10].
- This increase in charge transfer resistance increases the inhibition efficiency to 90.61 **%** in 1M HCl for 120mg/L concentration.
- C_{dl} decreases from **12.9** μF/cm^2 to **17** μF/cm^2 for blank HCl with the addition of compound **MON.**
- The change in double layer capacitance increases the surface coverage (θ) from **-1.04 - 0.31 in 1M HCl**

Table 5.1.3: Electrochemical impedance parameters for corrosion of mild steel at various concentrations of MON in 1M HCl

Conc mg/L	R_S Ω cm²	Y_0 μF/cm²	n	R_{ct} Ω cm²	IE %	fmax μHz	C_{dl} μF/cm²	θ	S^n	Icor (from Rct) μA/cm²	IE %
1M HCl											
Blank	6.81	8101	0.11	19		6.56	12.9		0.24	1238806	-
5	9.23	16579.2	-0.16	45	58.04	1.11	26.4	-1.0465	1.18	172156.7	61.07
15	16.6	16579.2	-0.18	68	72.42	0.75	23	-1.0465	1.8	111723.1	71.89
30	16.1	14444	-0.20	107	82.40	0.49	19.1	-0.7829	2.46	101198.7	82.6
45	19.2	11994.8	-0.20	114	83.39	0.51	18.2	-0.4806	2.17	82224.47	84.43
90	20.3	11429.6	-0.21	135	86.03	0.57	17.5	-0.4109	2.45	87238.72	86.68
120	22.5	10676	-0.23	201	**90.61**	0.66	17	-0.3178	3.41	72434.28	**90.87**

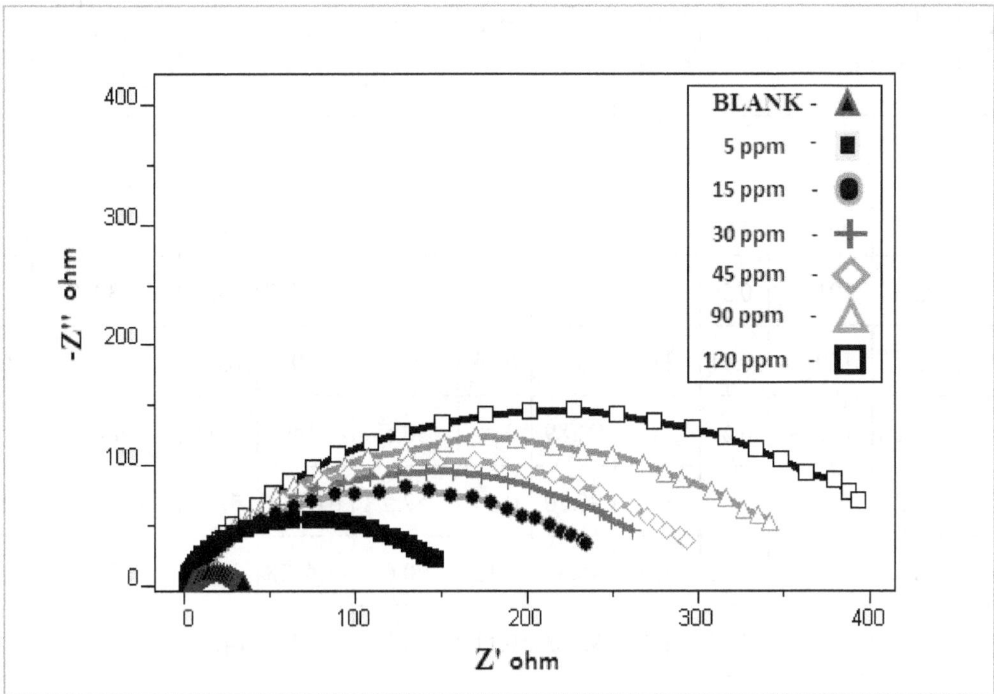

Figure 5.1.3. Nyquist curves recorded for mild steel electrode in 1M HCl Solution containing different concentrations of MON

5.1.4. PYR

The impedance behavior mild steel for compound **PYR Figure 5.1.4 & Table 5.1.4** in 1M HCl is discussed as follows:

- Analysis of the data reveals that, the R_{ct} values increase from **19** Ohmcm^2 for the blank HCl acid to **458** Ohmcm^2 with the addition of 120mg/L of compound PYR.
- Thereby increasing the IE from **92.24%** to **95.85%** in HCl.
- C_{dl} decreases from **12.9** to 10.2 μF/cm^2 for blank HCl at 120mg/L concentration.
- The decrease in C_{dl} values increases the surface coverage (θ) from -1.3 to -1.4 in HCl solution with the addition of compound **PYR.**
- The decrease in C_{dl} and increase in R_{ct} may be due to gradual replacement of water molecules by the adsorbed inhibitor.

Table 5.1.4: Electrochemical impedance parameters for corrosion of mild steel at various concentrations of compound PYR in 1M HCl

Conc mg/L	R_s Ω cm^2	Y_0 μF/cm^2	n	R_{ct} Ω cm^2	IE %	fmax μHz	C_{dl} μF/cm^2	θ	S^n	Icor (from Rct) μA/cm^2	IE %
1M HCl											
Blank	6.81	8076	0.25	19	-	6.52	12.9	-	0.24	1238806	-
5	9.23	19154	-0.56	245	92.24	0.21	31.8	-1.3717	7.47	283277	77.13
15	16.6	12943	-0.61	315	93.97	0.25	31.2	-0.6026	6.49	172594	86.07
30	16.1	17289	-0.61	372	94.89	0.16	30.5	-1.1407	1.02	108439	91.27
45	19.2	7511	-0.67	396	95.20	0.34	57.5	0.0700	4.74	125868	89.84
90	20.3	19970	-0.61	425	95.53	0.12	20.6	-1.4728	13.5	108030	91.28
120	22.5	19587	-0.62	458	**95.85**	0.11	10.2	-1.4253	14.29	65357	**94.72**

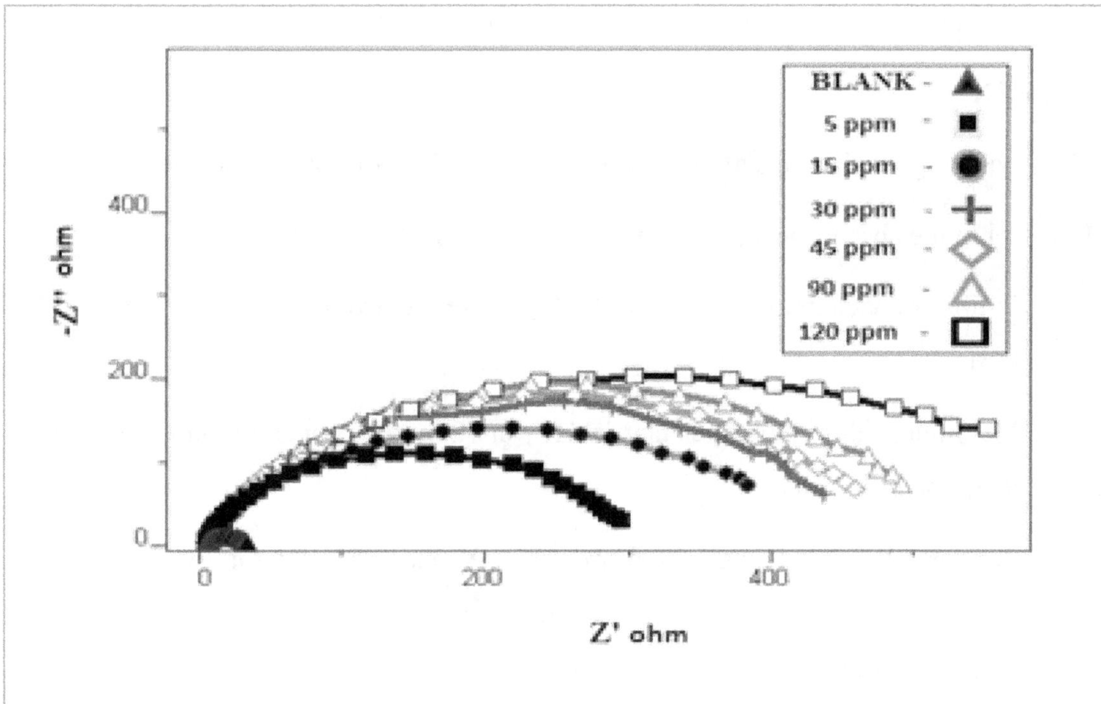

Figure 5.1.4. Nyquist curves recorded for mild steel electrode in 1M HCl Solution containing different concentrations of PYR

5.2. Potentiodynamic polarization studies

Potentiodynamic polarization experiments were undertaken to distinguish the effect of inhibitor on the anodic and cathodic corrosion reactions [11]–[14]. The Tafel polarization curves of mild steel in hydrochloric acid solution, in the absence and presence of different concentrations of the studied four drugs, are presented in **Figure 5.2.1 – 5.2.4** and their corresponding polarization parameters obtained by extrapolation of the linear Tafel curves to the corrosion current density (I_{corr}) are listed in **Table 5.2.1 – 5.2.4**

The polarization results obtained for the various drugs are discussed below.

5.2.1. BEX

BEX Analyzing the data in **Table 5.2.1** and **Figure 5.2**.1 polarization behavior of Mild steel in the absence and presence of compound **BEX** is summarized as follows:

- Inhibitor concentration increses, the I_{corr} decreases from 217 to 15 μA/cm^2 in HCl [15]
- Maximum inhibition efficiency in HCl at 120mg/L concentration is found to be **93.09%** respectively.
- Similarly E_{corr} value changes from -543 mV and -452 mV for 1M HCl with the addition of the inhibitors towards the noble direction. But there is no definite trend in the shift in the corrosion potential.
- With the addition of the inhibitor to the acid solution, both the anodic and cathodic tafel slopes are shifted in both the acid medium.

Table 5.2.1: Polarization parameters of mild steel at various concentrations of BEX in 1M HCl and corresponding inhibition efficiency values

Conc mg/L	Ecorr mV	Icorr μA/cm^2	IE %	ba mV/dec	bc mV/dec	R_p Ω cm^2	IE %	Icor (from Rp) μA /cm^2	IE %
1M HCl									
Blank	-543	217		101	117	108		63960	
5	-463	60	72.35	79	142	368	70.65	60486	25.43
15	-493	60	72.35	79	142	368	70.65	43473	32.03
30	-493	26	88.02	76	146	507	78.70	41898	34.49
45	-449	21	90.32	30	173	518	79.15	15702	75.45
90	-449	19	91.24	31	149	707	84.72	12422	77.10
120	-452	15	**93.09**	39	135	897	**87.96**	14647	**80.58**

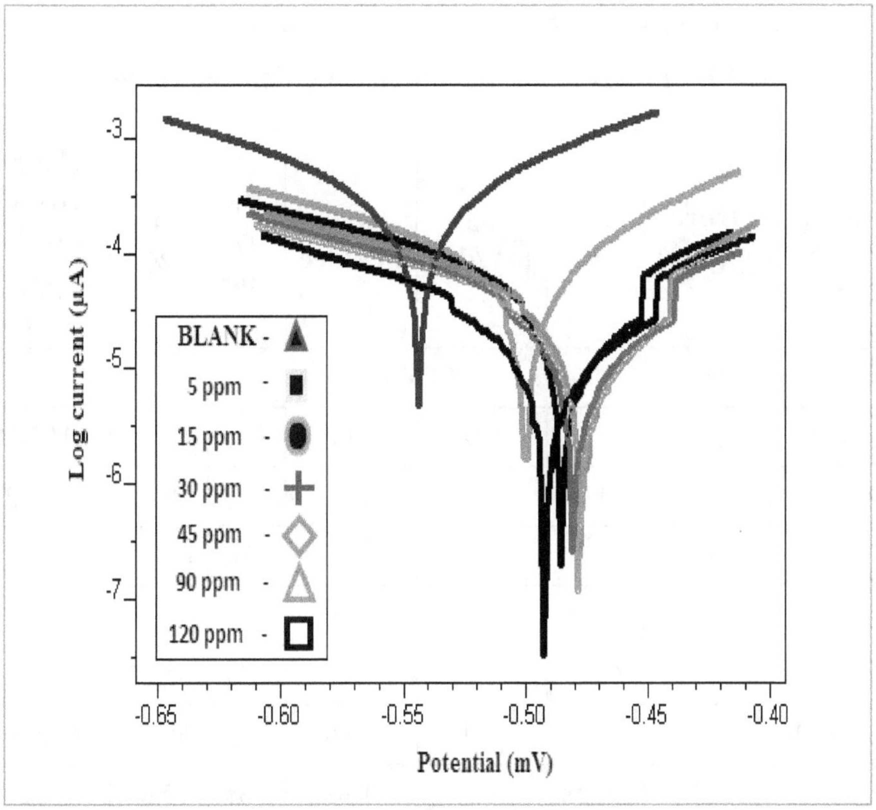

Figure 5.2.1. Polarization curves for mild steel in 1M HCl in the presence of compound BEX

5.2.2. CIL

It is found from **Table 5.2.2** and **Figure 5.2.2** with the addition of compound **CIL** to the acid medium, the polarization behavior is concise as follows:

- The I_{corr} decreases from **217 to 26** $\mu A/cm^2$ in HCl.
- This decrease in corrosion current leads to an increase in the inhibition efficiency with concentration to a maximum of **88.08%** in HCl
- Addition of the CIL also changes both the ba and bc values.
- E_{corr} value changes from -543 mV and -542 mV for 1M HCl with the addition of the inhibitors towards the noble direction [16]–[18].

Table 5.2.2 Polarization parameters of mild steel at various concentrations of CIL in 1M HCl corresponding inhibition efficiency values

Conc mg/L	Ecorr mV	Icorr $\mu A/cm^2$	IE %	ba mV/dec	bc mV/dec	R_p Ω cm^2	IE %	Icor (from Rp) μA $/cm^2$	IE %
1M HCl									
Blank	-543	217		101	117	108		110504	
5	-525	130	40.04	77	133	213	49.30	76722	20.57
15	-525	75	65.29	76	122	276	60.87	71346	35.44
30	-529	54	75.26	85	118	285	62.11	72480	45.41
45	-525	45	79.44	81	96	296	63.51	55293	78.96
90	-529	31	85.78	101	121	345	68.71	66214	82.08
120	-542	26	**88.08**	93	130	361	**70.08**	65211	**91.99**

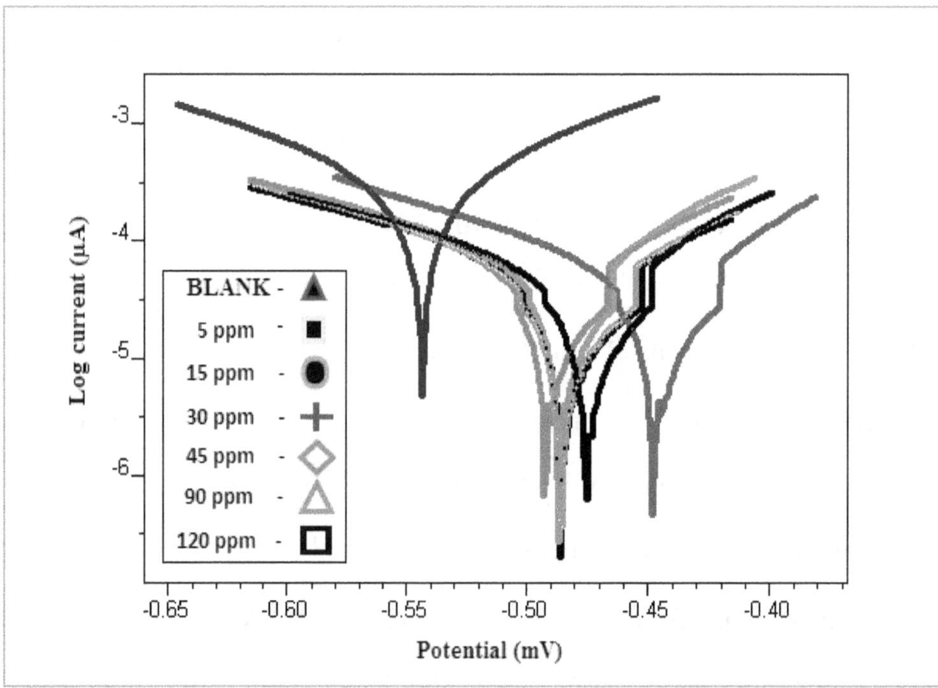

Figure 5.2.2 Polarization curves for mild steel in 1M HCl in the presence of compound CIL

5.2.3. MON

The analysis of the polarization results for compound **MON** given in **Table 5.2.3** and **Figure 5.2.3** shows that,

- The values of I_{corr} for mild steel in the inhibited solutions were lesser than those for the inhibitor-free solutions in both media.
- The inhibition efficiency increased with inhibitor concentration to a corresponding efficiency of **86.64%** respectively in 1 M HCl
- The addition of compound **MON** on acid corrosion of mild steel in both investigated acid medium, shift the E_{corr} values in the cathodic direction not more than 50mV which imply that the analyzed inhibitor acts as mixed type inhibitor and it influences both the metal dissolution and hydrogen evolution reaction [19]–[24].
- Both the anodic and cathodic tafel slopes are shifted in presence of the inhibitor in the acid medium.

Table 5.2.3 Polarization parameters of mild steel at various concentrations of MON in 1M HCl and corresponding inhibition efficiency values

Conc mg/L	Ecorr mV	Icorr dA/cm²	IE %	ba mV/dec	bc mV/dec	R_p Ω cm²	IE %	Icor (from Rp) µA/cm²	IE %
1M HCl									
Blank	-543	217	-	101	117	108	-	141791	-
5	-525	130	40.09	76	146	166	34.94	70888	50.01
15	-525	75	65.44	91	136	314	65.61	51258	63.85
30	-529	54	75.12	86	139	430	74.88	46275	67.36
45	-529	45	79.26	81	135	469	76.97	22564	84.09
90	-529	31	85.71	82	137	492	78.05	15160	89.31
120	-542	29	**86.64**	86	134	735	**85.31**	17876	**87.39**

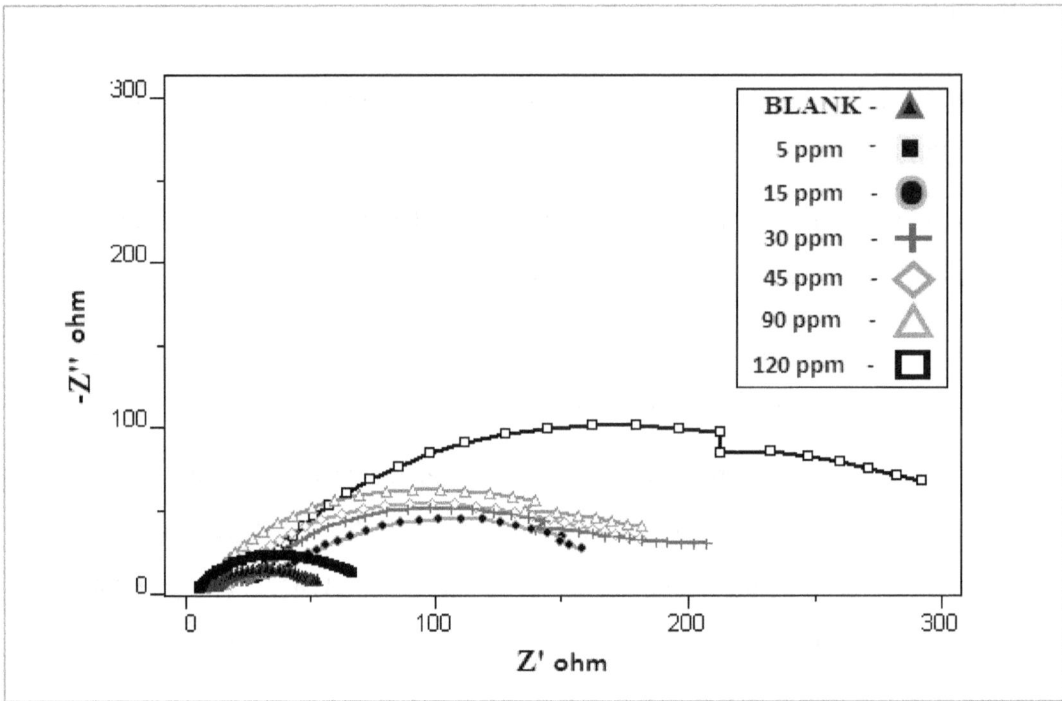

Figure 5.2.3. Polarization curves for mild steel in 1M HCl in the presence of compound MON

5.2.4. PYR

The polarization behavior of mild steel in the presence of the as given in **Table 5.2.4** and **Figure 5.2.4** is discussed below.

- The cathodic and anodic curves showed lower current density in the presence of the PYR derivative to a maximum of **88.08%** in 1M HCl
- The addition of the inhibitor changes the values of E_{corr} to the cathodic direction, so compound **PYR** can be considered as a slightly cathodic inhibitor.
- Tafel slopes b_a and b_c clearly indicate that both the cathodic and anodic reactions are inhibited thus reducing the corrosion on mild steel surface.

Table 5.2.4: Polarization parameters of mild steel at various concentrations of PYR in 1M HCl and corresponding inhibition efficiency values

Conc mg/L	Ecorr mV	Icorr μA/cm²	IE %	ba mV/dec	bc mV/dec	R_p Ω cm²	IE %	Icor (from Rp) μA/cm²	IE %
1M HCl									
Blank	-543	217	-	101	117	108	-	217938	-
5	-525	130	40.04	76	146	166	34.88	26447	87.86
15	-525	75	65.29	91	136	314	65.58	21978	89.92
30	-529	54	75.26	86	139	430	74.85	25614	88.25
45	-529	45	79.44	81	135	469	76.95	29956	86.25
90	-529	31	85.78	82	137	492	78.03	26090	88.03
120	-542	26	**88.08**	86	134	735	**85.30**	18584	**91.47**

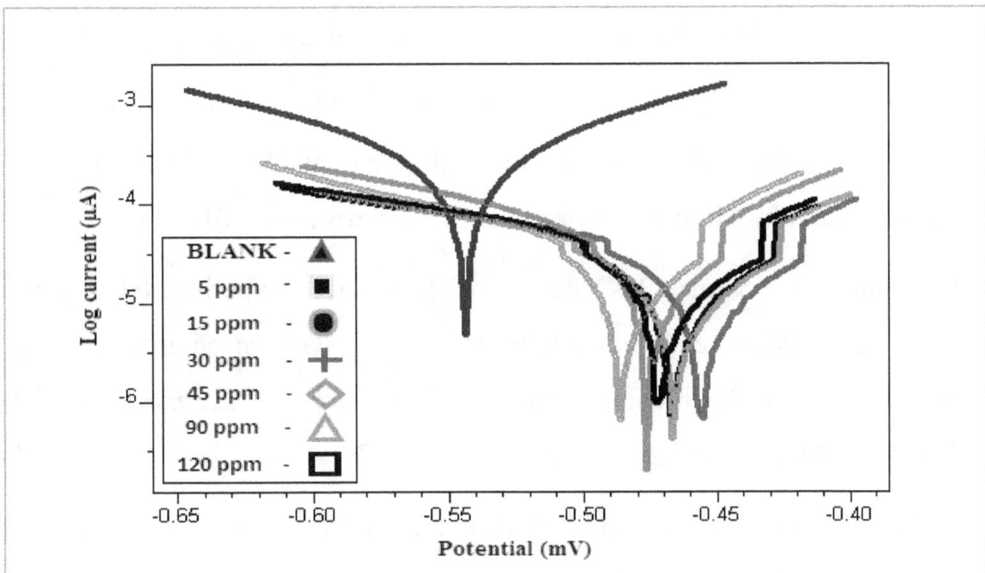

Figure 5.2.4 Polarization curves for mild steel in 1M HCl in the presence of compound PYR

References

[1] M. Ouakki *et al.*, "Quantum chemical and experimental evaluation of the inhibitory action of two imidazole derivatives on mild steel corrosion in sulphuric acid medium," *Heliyon*, vol. 5, no. 11, 2019.

[2] A. K. Singh *et al.*, "Evaluation of anti-corrosion performance of an expired semi synthetic antibiotic cefdinir for mild steel in 1 M HCl medium: An experimental and theoretical study," *Results Phys.*, vol. 14, 2019.

[3] A. Khadiri *et al.*, "Gravimetric, electrochemical and quantum chemical studies of some pyridazine derivatives as corrosion inhibitors for mild steel in 1 M HCl solution," *J. Taiwan Inst. Chem. Eng.*, vol. 58, pp. 552–564, 2016.

[4] P. Dohare, D. S. Chauhan, A. A. Sorour, and M. A. Quraishi, "DFT and experimental studies on the inhibition potentials of expired Tramadol drug on mild steel corrosion in hydrochloric acid," *Mater. Discov.*, vol. 9, pp. 30–41, 2017.

[5] Y. Koumya, R. Idouhli, A. Oukhrib, M. Khadiri, A. Abouelfida, and A. Benyaich, "Synthesis, Electrochemical, Thermodynamic, and Quantum Chemical Investigations of Amino Cadalene as a Corrosion Inhibitor for Stainless Steel Type 321 in Sulfuric Acid 1M," *Int. J. Electrochem.*, vol. 2020, pp. 1–10, 2020.

[6] P. Arockiasamy, X. Q. R. Sheela, G. Thenmozhi, M. Franco, J. W. Sahayaraj, and R. J. Santhi, "Evaluation of Corrosion Inhibition of Mild Steel in 1 M Hydrochloric Acid Solution by Mollugo cerviana," *Int. J. Corros.*, vol. 2014, 2014.

[7] S. Kumar, D. Sharma, P. Yadav, and M. Yadav, "Experimental and quantum chemical studies on corrosion inhibition effect of synthesized organic compounds on N80 steel in hydrochloric acid," in *Industrial and Engineering Chemistry Research*, 2013, vol. 52, no. 39, pp. 14019–14029.

[8] A. E. A. S. Fouda, S. A. Abd El-Maksoud, E. H. El-Sayed, H. A. Elbaz, and A. S. Abousalem, "Experimental and surface morphological studies of corrosion inhibition on carbon steel in HCl solution using some new hydrazide derivatives," *RSC Adv.*, vol. 11, no. 22, pp. 13497–13512, 2021.

[9] M. Mobin, S. Zehra, and R. Aslam, "L-Phenylalanine methyl ester hydrochloride as a green corrosion inhibitor for mild steel in hydrochloric acid solution and the effect of surfactant additive," *RSC Adv.*, vol. 6, no. 7, pp. 5890–5902, 2016.

[10] F. E. Abeng, V. C. Anadebe, V. D. Idim, and M. M. Edim, "Anti-corrosion behaviour of expired tobramycin drug on carbon steel in acidic medium," *South African J. Chem.*, vol. 73, no. 1, pp. 125–130, 2020.

[11] E. M. Attia, N. S. Hassan, and A. M. Hyba, "Corrosion Protection of Tin in 1M HCl by Expired Novacid Drug-Part I," *Int. J. Adv. Res.*, vol. 4, no. 2, pp. 872–886, 2016.

[12] C. Verma, M. A. Quraishi, and A. Singh, "5-Substituted 1H-tetrazoles as effective corrosion inhibitors for mild steel in 1 M hydrochloric acid," *J. Taibah Univ. Sci.*, vol. 10, no. 5, pp. 718–733, 2016.

[13] A. H. Seikh and E. S. M. Sherif, "Effects of immersion time and temperature on the corrosion of API 5L Grade X-65 Steel in 1.0 M H2SO4 pickling solution," *Int. J. Electrochem. Sci.*, vol. 10, no. 1, pp. 895–908, 2015.

[14] M. Marinescu, "Recent advances in the use of benzimidazoles as corrosion inhibitors," *BMC Chemistry*, vol. 13, no. 1. BioMed Central Ltd., 2019.

[15] A. Pradityana, Sulistijono, and D. A. Shahab, "Application of myrmecodia pendans extract as a green corrosion inhibitor for mild steel in 3.5% NaCl solution," in *Applied Mechanics and Materials*, 2014, vol. 493, pp. 684–690.

[16] E. M. Attia, "Expired Farcolin Drug as Corrosion Inhibitor for Carbon Steel in 1M HCl Solution," *J. Basic Appl. Chem.*, vol. 5, no. 1, pp. 1–15, 2015.

[17] E. M. Attia, "Expired Farcolin Drug as Corrosion Inhibitor for Carbon Steel in 1MHCl Solution," 2015.

[18] R. S. Abdel Hameed, E. A. Ismail, H. I. Al-Shafey, and M. A. Abbas, "Expired Indomethacin Therapeutics as Corrosion Inhibitors for Carbon Steel in 1.0 M Hydrochloric Acid Media," *J. Bio- Tribo-Corrosion*, vol. 6, no. 4, 2020.

[19] N. Raghavendra, "Expired abacavir sulfate drug as non-toxic corrosion inhibitor for mild steel (MS) in 3 M hydrochloric acid system," *Gazi Univ. J. Sci.*, vol. 32, no. 4, pp. 1113–1121, 2019.

[20] Z. Zhang *et al.*, "Molecule adsorption and corrosion mechanism of steel under protection of inhibitor in a simulated concrete solution with 3.5% NaCl," *RSC Adv.*, vol. 8, no. 37, pp. 20648–20654, 2018.

[21] A. Srinivasulu and P. K. Kasthuri, "Study of inhibition and adsorption properties of mild steel corrosion by expired pharmaceutical Gentamicin drug in hydrochloric acid media," *Orient. J. Chem.*, vol. 33, no. 5, pp. 2616–2624, 2017.

[22] G. K. Shamnamol, K. P. Sreelakshmi, G. Ajith, and J. M. Jacob, "Effective utilization of drugs as green corrosion inhibitor-A review," *AIP Conf. Proc.*, vol. 2225, pp. 1–5, 2020.

[23] N. O. Eddy, P. O. Ameh, and N. B. Essien, "Experimental and computational chemistry studies on the inhibition of aluminium and mild steel in 0.1 M HCl by 3-nitrobenzoic acid," *J. Taibah Univ. Sci.*, vol. 12, no. 5, pp. 545–556, 2018.

[24] X. Xu, A. Singh, Z. Sun, K. R. Ansari, and Y. Lin, "Theoretical, thermodynamic and electrochemical analysis of biotin drug as an impending corrosion inhibitor for mild steel in 15% hydrochloric acid," *R. Soc. Open Sci.*, vol. 4, no. 12, 2017.

Chapter VI

CHAPTER VI

THEORETICAL INVESTIGATION ON THE INHIBITIVE PROPERTY OF THE STUDIED DRUGS

Study on corrosion inhibition plays an important role in the field of research. Quantum chemical methods have been used to examine the characteristics of the inhibitor /surface mechanism and to describe the structural nature of the inhibitor in the corrosion process [1]–[3]. Theoretical parameters are widely used to characterize the molecular structure of the inhibitors and their adsorption mechanism on the metal surface [4 - 5]. Two main advantages of using theoretical parameters are (i) the proposed mechanism for the studied compounds could be accounted in terms of chemical reactivity and (ii) the substituents present in the compounds can be directly characterized based on molecular structure. Some heterocyclic organic drugs are used as corrosion inhibitors for mild steel in acid environments [6 - 9]. The aim of this present study is to investigate the inhibitive nature of the drugs namely, BEX, CIL, MON and PYR by using quantum chemical calculations.

6.1. Structure of the inhibitor

The chemical structure of the four inhibitor systems is shown the following table.

Structure of the compound	
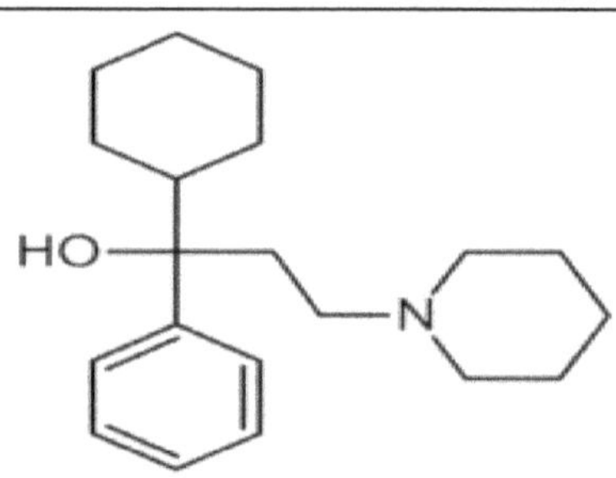 **BEX**	**MON** 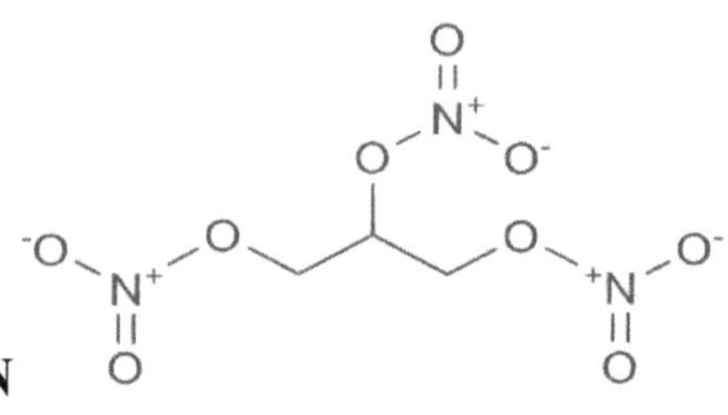
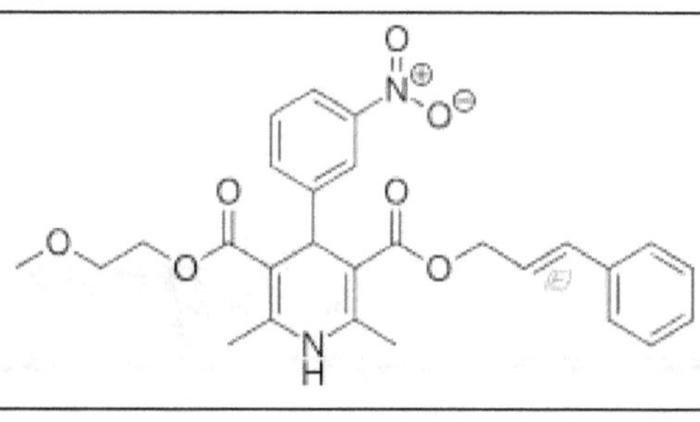 **CIL**	**PYR** 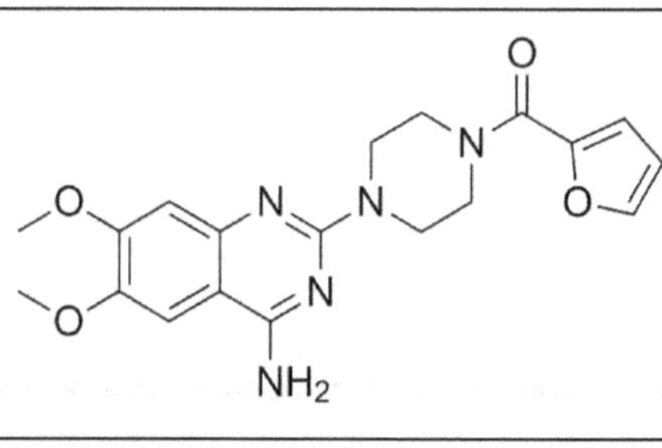

BEX

The structure of BEX contains a piperidine ring, cyclohexyl ring, phenyl ring and a hydroxyl group.

CIL

The drug **CIL** contains dihydropyridine ring with two carboxylate substituents at meta position, two methyl group at ortho position and a nitrophenyl ring at the para position.

MON

The drug MON is an aliphatic compound, glyceryl trinitrate with three nitrate group.

PRAZOPILL

It is an N-heterocyclic ring system an amino group containing a quinazoline ring with a two methoxy groups, are as ring substituents. It also contains a piperazine ring and furan ring.

The presence of hetero atoms and aromatic rings in the molecule facilitate them to get adsorbed on the metal surface, thereby inhibiting the corrosion process[10]–[13]. Thus the structure of these four inhibitors is optimized for their minimum energy configuration, to study its molecular property.

6.2. Optimized structure of the non-protonated species

Figure 6.1 reports the lowest energy conformer corresponding to each structure (optimized geometry) which is discussed as follows: The C-C bond lengths (Å) of the aromatic ring are in the range 1.39 – 1.41 for all the studied compounds[14]–[16]. These ranges are comparable to the range of the calculated and experimental bond lengths in isolated benzene ring or in an aliphatic system[17 - 19]. Except MON all the other three drugs where expected to have a non-planar structure, because of its bulky groups and bigger structure [20 - 21].

BEX having a cyclohexane, a piperidine, a phenyl ring and CIL with different bulky substituents is completely non-planar. The aliphatic compound MON is planar with all the three nitrate groups in the same plane. In PYR the piperazine and also the quiazoline rings are in the same plane with the furan ring perpendicular to this plane.

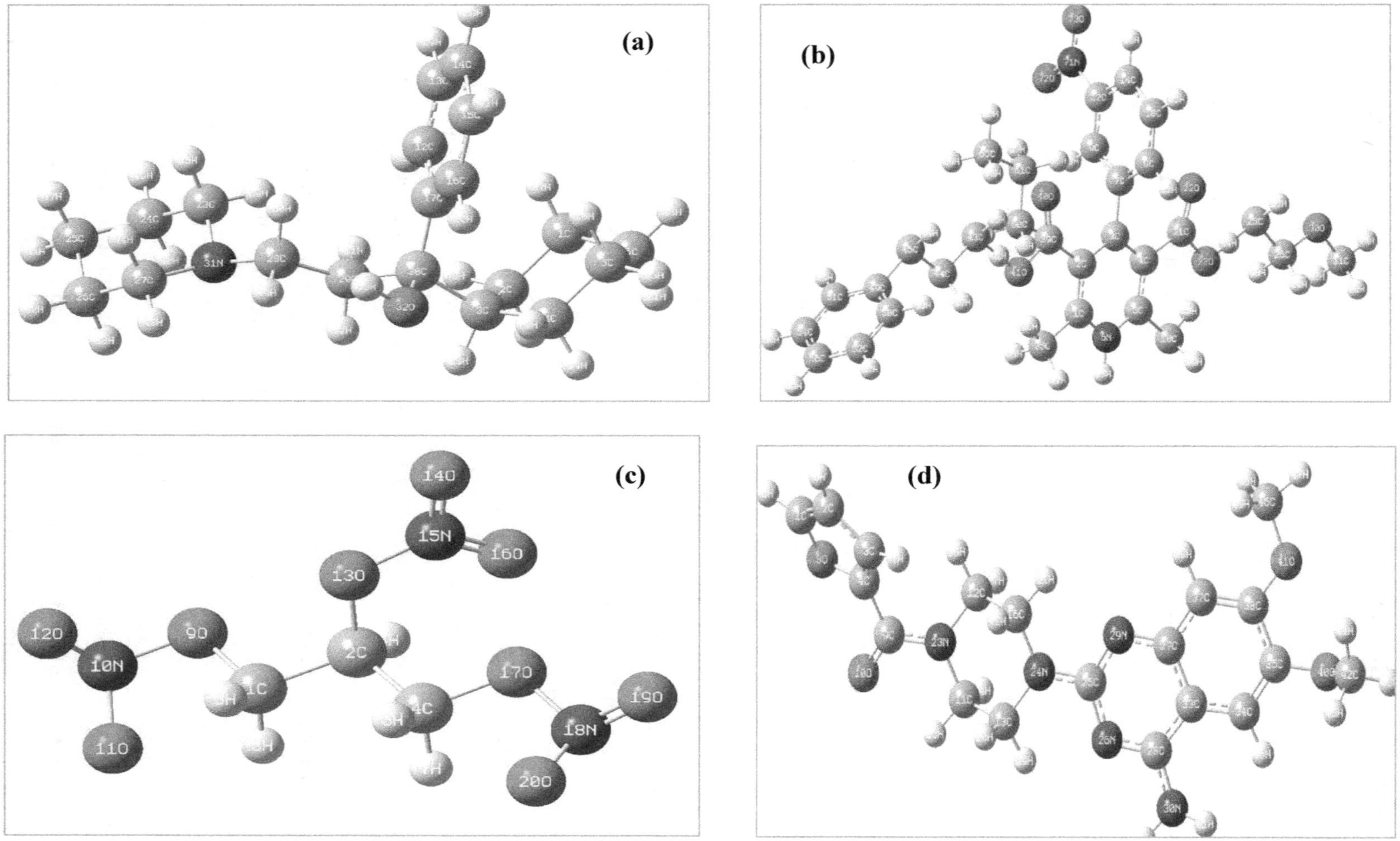

Figure 6.1 Optimized geometry of (a) BEX (b) CIL (c) MON (d) PYR

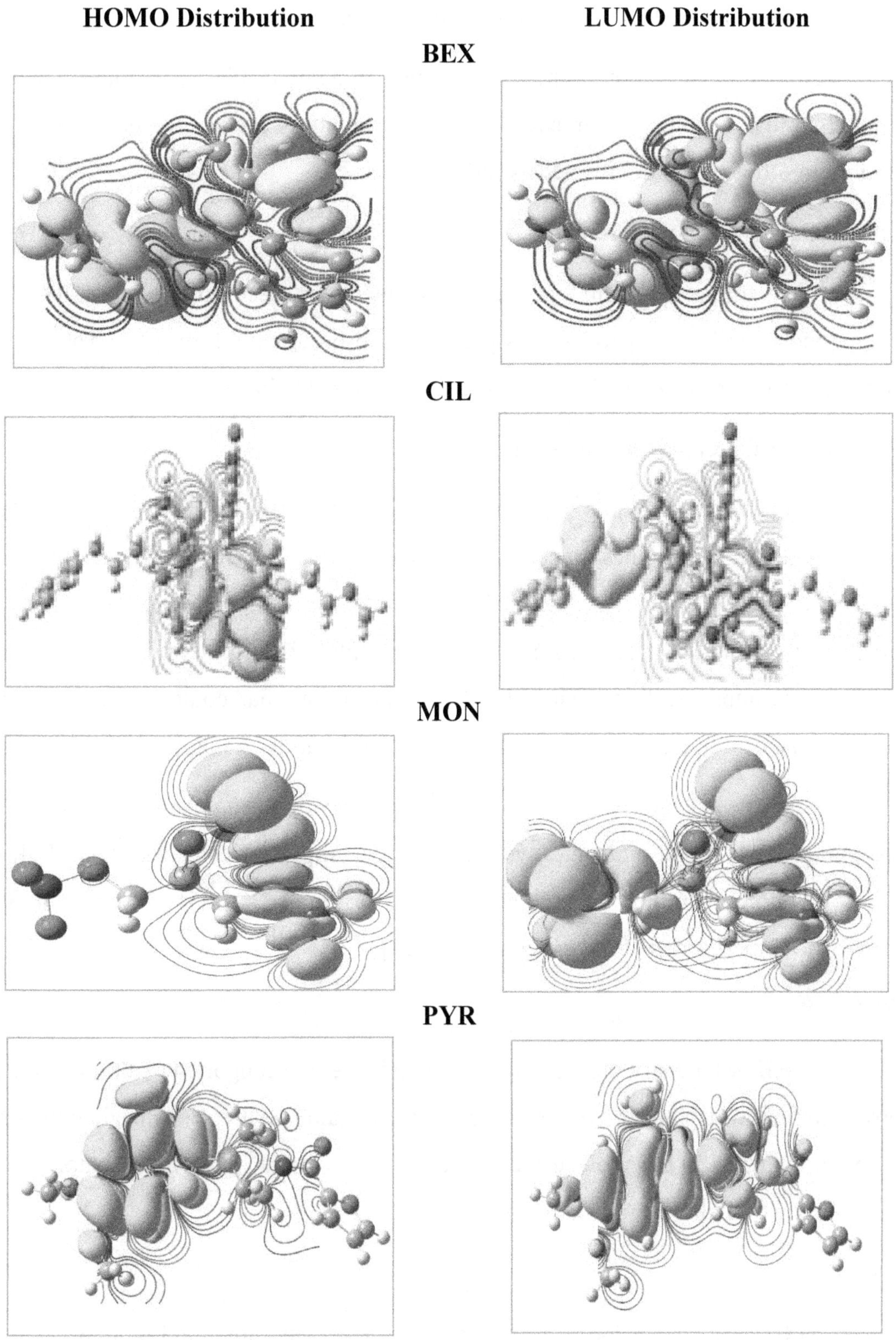

Figure 6.2 FMO distribution of the studied drug

6.3 Molecular properties of the non-protonated species

Non protonated form of the studied drugs of the molecular electronic structures are related to the quantum chemical parameters viewing excellent association with the inhibition efficiency is deliberated as follows

Global reactivity parameters

The various quantum chemical parameters are used to designate the studied compound inhibitive property in gas phase of non-protonated form [22]. The important parameters are presented in **Table 6.1** and are discussed below.

i) Frontier molecular orbital energy

According to Fukui's frontier molecular orbital theory, the frontier molecular orbital together with highest occupied molecular orbital HOMO and lowest unoccupied molecular orbital LUMO related to the inhibitor molecules adsorption centers. The HOMO is the orbital that act as the electron donor, meanwhile it is the outermost (highest energy) orbital containing electrons. The LUMO is the orbital that could act as the electron acceptor, since it is the innermost (lowest energy) orbital that act as the electron acceptor.

Highest values of the E_{HOMO} reflect the tendency of the molecule to donate electrons to appropriate acceptor molecules with low energy, empty molecular orbitals. High value of E_{HOMO} simplify Inhibition efficiency and adsorption [23 – 25]. Lower value of the E_{LUMO}, accept electrons of the molecule [26 - 27].

BEX

The HOMO distribution is over the **piperidine** ring supported by the six membered aromatic ring, which act as binding centers to the metal. LUMO is localized on the **phenyl ring** through which the molecule can accept the electrons from the Fe atom with its antibonding orbital to form a back-donating bond.

CIL

In **CIL**, HOMO is localized on the pyridine ring as shown in Figure **6.2** which act as binding sites to the metal surface. The LUMO is distributed over the dicarboxylate part supported by the aromatic ring to form a back-donating bond.

Table 6.1. Quantum chemical parameters obtained by DFT calculation using Gaussian 09 for the studied four drugs

Quantum chemical Parameters	BEX	CIL	MON	PYR
Molecular Formula	$C_{20}H_{31}No$	$C_{27}H_{28}N_2O_7$	$C_{20}H_{31}No$	$C_{19}H_{21}N_5O_4$
Total Energy (amu)	-855.52	-1776	-947.36	-1311.33
E_{HOMO} (amu)	-4.84	-5.52	-11.99	-5.156
E_{LUMO} (amu)	-2.13	-0.414	-1.80	-0.8517
ΔE (amu)	2.71	5.9	10.188	4.3
Dipole Moment (μ)	10.177	12.83	4.6771	5.1879
Global Hardness (η)	0.0497	2.55	5.09	2.15
Global Softness (σ)	20.12	0.19	0.098	0.232
Electronegativity (χ)	0.1281	2.967	6.89	3
ΔN	0.21	0.8	0.1	1

MON

Both the HOMO and LUMO distribution in MON is found on all the three nitrate groups indicating the electron donating and also the electron accepting ability of heteroatoms (N and O) to the metal atom[28]. The vacant d-orbital of Fe atom have the tendency to accept the electron from HOMO of the drug to form a coordinate bond. The LUMO orbitals enable the molecule could accept electrons from metal atom to form a back-donating bond with its anti-bonding orbitals.

PYR

In PYR, HOMO is localized over the entire molecule except furyl ring. The localization is entirely over heteroatoms indicating that these are the active sites through which they donate electrons to unoccupied orbitals of Fe atoms on the metal surface. The LUMO is localised over the quinazoline ring with the methoxy substituents which are expected to accept electrons from the metal atom.

ii) Energy gap (ΔE)

A different important parameter of quantum chemical studies of energy gap (**ΔE**), which indicates that inhibitor molecule adsorption comportment and the reactivity of the molecules. Energy gap value is low that exhibits inhibitor molecule will be easily adsorbed on the metal surface [29 - 30].

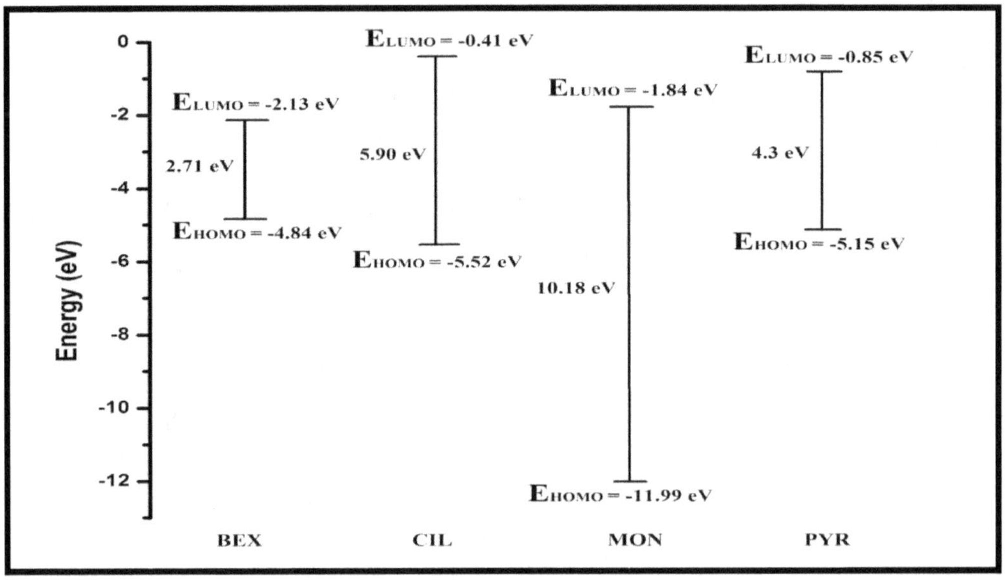

Figure 6.3 Schematic diagram of frontier molecular orbitals for drugs with their ΔE values

Figure 6.3 shows the correlation diagram of frontier molecular orbitals for investigated drugs with their calculated energy gap. As the energy gap value decreases, the chemical reactivity of the drugs increases resulting in the increase of inhibition efficiency. The energy level diagram shows BEX has lowest energy gap among all the drugs. Hence it is expected to show a greater inhibitive action than other investigated drugs. The energy gap values follow the order

BEX < PYR < CIL < MON

Together with the energy gap the other properties and also its optimized structure helps in predicting the adsorption ability of the studied molecules.

iii) Dipole moment (μ)

The dipole moment (**μ**) is another key that can be used to calculate the path of the corrosion inhibition process and it is related to the electron distribution in a molecule and also the magnitude the polarity in a bond [31 - 32]. While literature is varying the use of dipole moment is a direction on the inhibition process of corrosion. Generally that the polar compound having adsorption on the metal surface having high dipole moment must lead to the well inhibition efficiency. The dipole moment is the additional indicator, to calculate the number of electrons distributed within a molecule.

The higher the value of dipole moment, the greater is the adsorption of polar groups on the surface of the metal leading to better inhibition efficiency. The trend in dipole moment for the studied drugs is as follows:

CIL (12.83) > BEX (10.17) > PYR (5.18) > MON (4.67)

In general, a more polar compound is assumed to have higher charge separation resulting in better adsorption. Also, this trend is found to be in correlation with the results derived from other quantum chemical indices.

iv) Ionization potential (IP) and electron affinity (EA)

E_{HOMO} is directly related to ionization potential which specifies high reactivity of the atoms and molecules and E_{LUMO} is related to electron affinity [33].

Inhibition efficiency values of the investigated drugs increases with increase in the values of ionization energy but decreases with decrease in the values of electron affinity. Furthermore, E_{HOMO} is directly related to the Ionization potential while, E_{LUMO} is directly related to electron affinity.

v) Softness (σ) and Hardness (η)

Softness **(σ)** & Hardness **(η)** is the next important parameters of quantum chemical studies. Softness is interprets, the inhibitor molecules ability of the adsorption studies. On the basis of Activation hardness is energy gap. Polarizability, is closely related to the parameter of Hardness[34]. Decrease the value of energy gap indications to the easily polarize the molecule.

Hardness **(η)** and softness **(σ)** is used to study the inhibitor molecules reactivity. According to HSAB, hard and soft acid/base, soft acid to prefer to coordinate soft base, hard acid to prefer to coordinate hard bases. Maximum softness of the inhibitor molecule will coordinate to the soft acid of Fe and low energy gap.

Compared to the hard ones, soft molecules are easily more reactive due to its ability to accept the electrons easily. It is obvious from the literature that larger the energy gap, lower is the chemical reactivity as the energy gap is related to both the global softness and absolute hardness of a compound. Among the studied drugs, the trend in global softness is provided as follows:

BEX (20.12) > **PYR** (0.232) > **CIL** (0.19) > **MON** (0.098)

The trend in softness shows that BEX has more soft nature when compared to other drugs with similar trend in the energy gap. Hence, this drug is expected to show better inhibition efficiency.

vi) Electronegativity

Electronegativity (χ) is the most important chemical property that describes the ability of a molecule to attract electrons in a covalent bond with itself. The derived result indicates that the electron flow is from the molecule with the low electronegativity to a higher electronegativity. When metal and inhibitor come into contact, the electron flows from the lower χ (inhibitor) to the upper χ (metal) until the chemical potentials equalize. In this present study, MON is having a greater electronegativity than BEX, CIL, and PYR.

Local selectivity parameters

The local charge on the atoms in a compound plays a vital role in studying various chemical reactions and physicochemical properties. Mulliken charge population analysis is mainly employed for the evaluation of charge distribution in a compound[35]. Further, the molecular polarity can be described by atomic charges.

Mulliken Charge Density

The electron densities on the atoms provide data on the charge distribution and hence the chemical reactivity of the compound. It also gives information on the selectivity i.e., the specific sites in which a certain type of reaction could occur. Sites with the highest negative charges are susceptible to an electrophilic attack. **Table 6.2 – Table 6.5** reveals the Mulliken charge distribution of non-hydrogen atoms of the investigated drugs studied drugs using B3LYP/6-31 G (d,p) method.

Table 6.2. Calculated Fukui indices, Mulliken charge density and local softness for the BEXOL

Atom Label	q_N	f^-	f^+	S^-	S^+
1C	-0.42	0.00	0.00	0.00	0.00
2C	-0.41	0.00	0.02	0.00	0.01
3C	-0.22	0.07	0.01	0.03	0.00
4C	-0.66	-0.01	0.03	0.00	0.02
5C	-0.43	-0.02	0.00	-0.01	0.00
6C	-0.41	0.01	0.01	0.00	0.00
30C	0.21	-0.04	0.01	-0.02	0.00
12C	-0.24	-0.49	-0.07	-0.24	-0.04
17C	-0.07	-0.13	-0.10	-0.07	-0.05
32O	-0.69	-0.52	0.52	-0.25	0.25
16C	-0.24	-0.03	0.00	-0.01	0.00
13C	-0.24	-0.24	0.24	-0.12	0.12
15C	-0.23	0.00	0.03	0.00	0.01
14C	-0.23	0.00	-0.06	0.00	-0.03
29C	-0.24	-0.01	-0.05	-0.01	-0.02
28C	-0.18	-0.18	0.18	-0.09	0.09
31N	-0.09	0.60	-0.38	0.29	-0.19
7C	-0.084	0.00	0.00	0.00	0.00
24C	-0.41	-0.17	0.00	-0.08	0.00
25C	-0.41	-0.41	0.41	-0.20	0.20
26C	-0.41	0.02	0.00	0.01	0.00

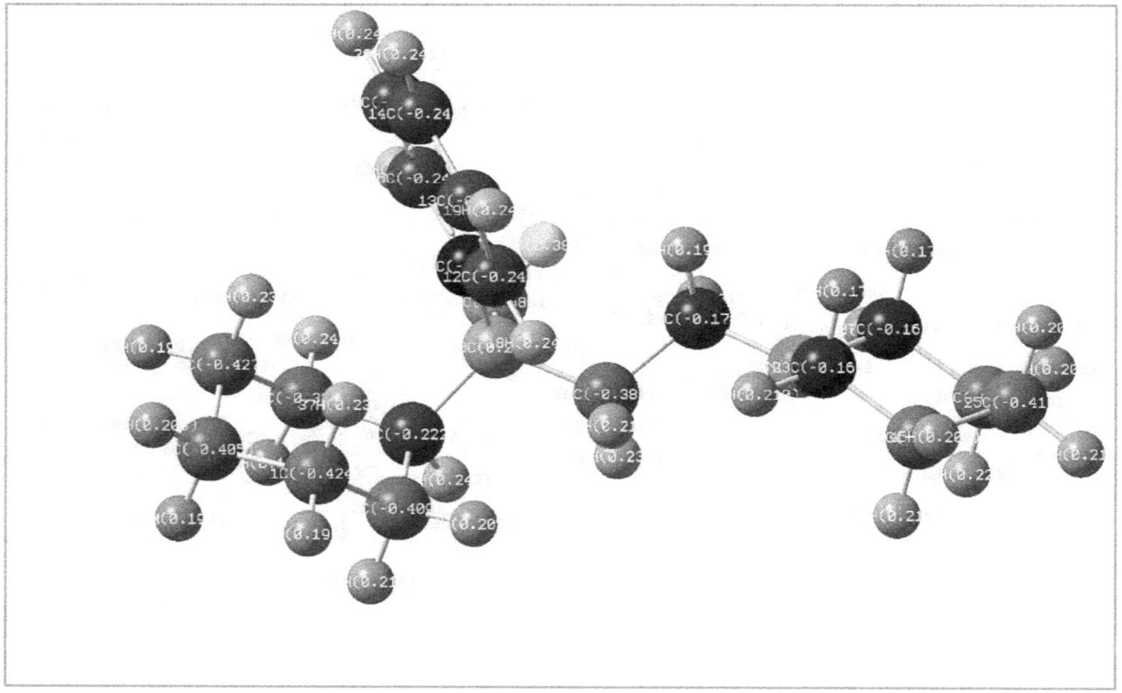

Figure 6.4. Mulliken charge distribution among the different atoms in the drug BEX

Table 6.3. Calculated Fukui indices, Mulliken charge density and local softness for the CILNY

Atom Label	q_N	f^-	f^+	S^-	S^+
1C	-0.19	-0.02	-0.10	-0.01	-0.05
2C	0.28	-0.02	0.00	-0.01	0.00
3C	-0.17	-0.01	-0.08	-0.01	-0.04
4C	-0.11	-0.06	0.09	-0.03	0.05
5C	-0.13	0.04	-0.13	0.02	-0.07
46O	-0.40	0.02	-0.01	0.01	0.00
44N	0.18	-0.07	0.04	-0.03	0.02
45O	-0.39	-0.64	0.08	-0.31	0.04
11C	0.10	-0.06	-0.14	-0.03	-0.07
13C	-0.32	0.04	0.01	0.02	0.01
15C	0.25	0.62	-0.04	0.30	-0.02
43N	-0.97	0.00	0.03	0.00	0.01
14C	0.25	-0.12	-0.03	-0.06	-0.01
12C	-0.31	-0.08	-0.05	-0.04	-0.03
31C	-0.11	-0.01	-0.18	-0.01	-0.09
32C	-0.25	0.02	0.02	0.01	0.01
33C	-0.17	-0.03	-0.02	-0.02	-0.01
24C	-0.09	0.02	0.01	0.01	0.00
23C	-0.22	0.00	-0.01	0.00	-0.01
22C	-0.23	0.00	0.00	0.00	0.00
34C	0.98	-0.06	-1.39	-0.03	-0.68
41O	-0.75	0.00	0.00	0.00	0.00
39O	-0.74	0.01	0.00	0.01	0.00
35C	-0.10	0.02	0.02	0.01	0.01
36C	0.08	0.15	-0.16	0.07	-0.08
38O	-0.65	-1.31	0.00	-0.64	0.00
46C	0.25	0.25	-0.25	0.12	-0.12

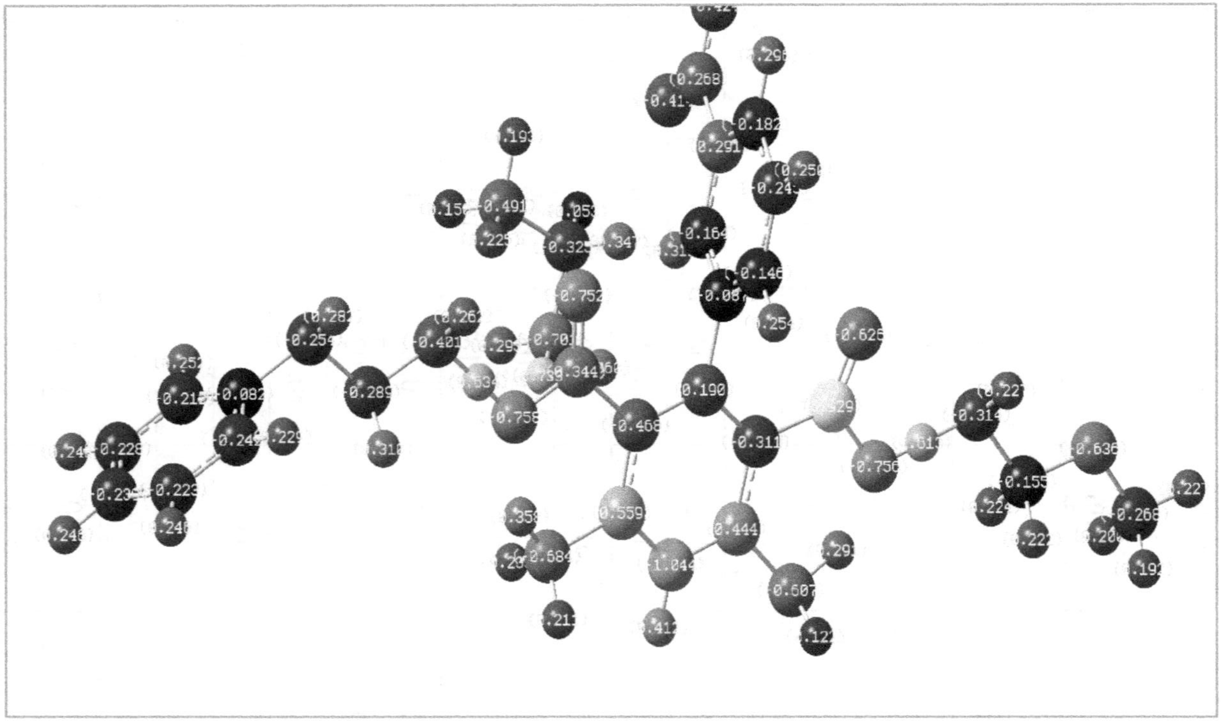

Figure 6.5. Mulliken charge distribution among the different atoms in the drug CIL

Table 6.4 Calculated Fukui indices, Mulliken charge density and local softness for the MONIT

Atom Label	q_N	f^-	f^+	S^-	S^+
1C	-0.09	-0.02	-0.02	-0.01	-0.01
2C	-0.01	0.39	-0.38	0.19	-0.19
3C	-0.08	0.07	0.06	0.03	0.03
4O	-0.45	-0.08	-0.05	-0.04	-0.02
13N	0.56	0.07	-0.10	0.03	-0.05
9O	-0.31	0.04	-0.12	0.02	-0.06
10O	-0.22	-0.18	0.00	-0.09	0.00
5O	-0.38	-0.02	-0.04	-0.01	-0.02
15N	0.64	-0.06	-0.15	-0.03	-0.07
12O	-0.39	-0.20	-0.59	-0.10	-0.29
11O	-0.32	-0.19	-0.53	-0.09	-0.26
6O	-0.44	0.00	0.01	0.00	0.00
14N	0.64	0.00	0.00	0.00	0.00
7O	-0.30	-0.04	-0.04	-0.02	-0.02
8O	-0.33	-0.07	-0.02	-0.03	-0.01

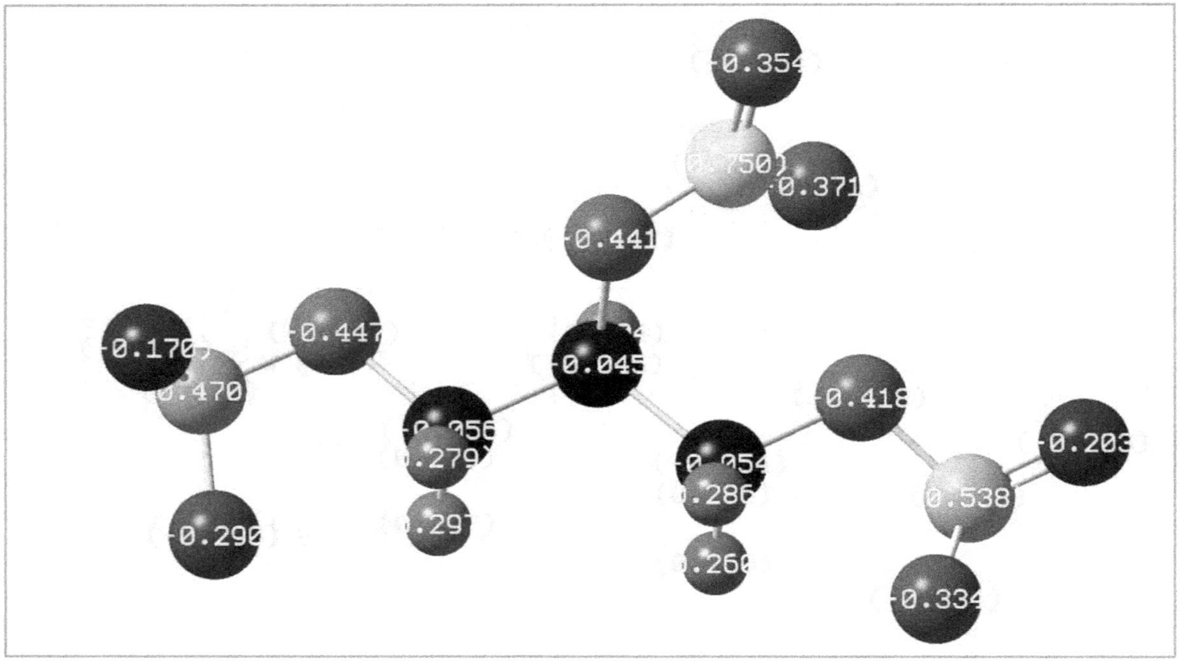

Figure 6.6. Mulliken charge distribution among the different atoms in the drug MON

Table 6.5. Calculated Fukui indices, Mulliken charge density and local softness for the PYR

Atom Label	q_N	f^-	f^+	S^-	S^+
1C	0.14	0.54	-0.33	0.27	-0.16
2O	-0.65	-0.07	-0.03	-0.04	-0.01
3C	-0.10	0.10	0.12	0.05	0.06
4C	-0.17	-0.16	-0.19	-0.08	-0.09
5C	-0.33	0.06	0.09	0.03	0.04
10C	0.94	-0.06	-0.04	-0.03	-0.02
32O	-0.64	-0.03	-0.07	-0.01	-0.03
25N	-0.89	-0.01	0.00	-0.01	0.00
13C	-0.17	0.02	-0.03	0.01	-0.01
14C	-0.16	-0.29	-0.03	-0.14	-0.01
26N	-0.94	0.00	0.00	0.00	0.00
11C	-0.16	-0.01	0.00	-0.01	0.00
15C	-0.18	0.02	0.02	0.01	0.01
27N	1.13	0.01	0.00	0.00	0.00
28N	-0.24	-0.01	-0.05	-0.01	-0.02
16C	-0.88	0.04	0.01	0.02	0.01
29N	0.92	-0.01	-0.01	0.00	0.00
18C	-0.18	-0.01	-0.01	0.00	0.00
19C	0.38	0.01	0.01	0.00	0.01
20C	-0.27	0.00	0.00	0.00	0.00
21C	0.43	-0.01	-0.01	-0.01	-0.01
22C	0.34	-0.02	-0.02	-0.01	-0.01
30O	-0.20	0.00	0.00	0.00	0.00
23C	-0.73	-1.47	-0.01	-0.72	0.00
31O	-0.27	0.01	0.00	0.01	0.00
24C	-0.70	0.00	0.00	0.00	0.00

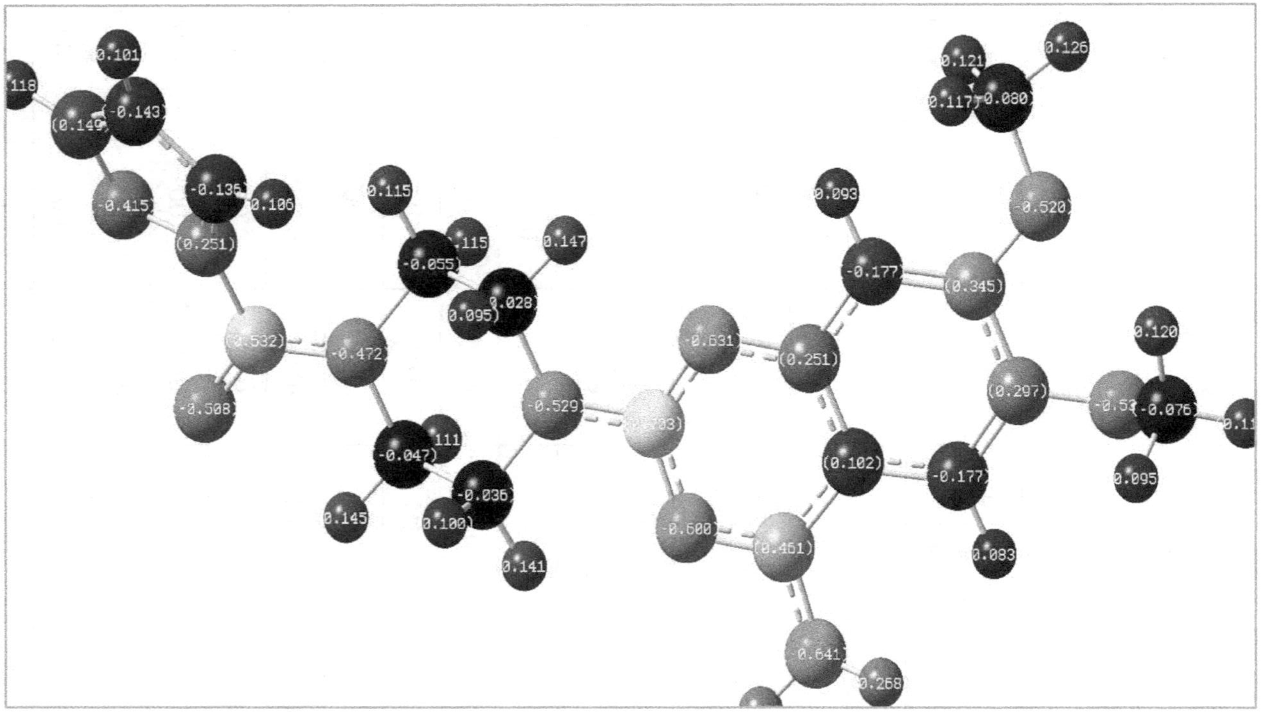

Figure 6.7. Mulliken charge distribution among the different atoms in the drug PYR

BEXOL

Table 6.2 shows that in BEX, 31N atom carries the highest negative charge. Hence, this atom is susceptible to an electrophilic attack in which this will readily donate its electrons to the electrophilic species. When metal surface acts as electrophilic species, the donated electrons goes to the vacant or partially filled d-orbitals of the steel substrate allowing the compound to get adsorbed on the surface of the metal (adsorption phenomena)[36]. The results coincides with the electron density potential surface indicating that N atoms are susceptible to an electrophilic attack. All C atoms that are directly bonded to N atoms are electron deficient and therefore it could be subjected to nucleophilic attack. The aromatic ring has a delocalization of negative charge on all its carbon atoms suggesting that N atom is the possible site that interact with the metal surface resulting in a protective layer formation on the metal surface.

CILNY

Figure 6.3 represents Milliken charge distribution calculated on optimized geometry among the non-hydrogen atoms using B3LYP/6-31 G (d,p) method. The results indicate that O atom carries negative charges than that of N atom. These atoms take part in electrophilic attack in which this will readily donate its electrons to the electrophilic species. The aromatic ring has a delocalization of negative charge on all its carbon atoms suggesting that O and N atoms are the possible sites that interact with the metal surface resulting in a protective layer formation on the metal surface.

MONIT

Mulliken charge distribution among the non-hydrogen atoms of optimized geometry of **Monit** is shown in **Figure 6.7** and the values are tabulated in **Table 6.4** The results show that the number of negatively charged atoms is more than those in other drugs. **N atoms** carry the highest negative charge. These atoms are therefore likely to readily donate electrons to the partially filled or vacant d orbitals of the metal and gets adsorbed on the metal surfaces to form a protective layer.

PRAZOPILL

Mulliken charge distribution among the non-hydrogen atoms of Prazopill is shown in **Figure 6.10** and the values are tabulated in **Table 6.5.** The results show that the number of negatively charged atoms is more than those in other drugs. **N and O atoms** carry the highest negative charges. These atoms are therefore likely to readily donate electrons to the partially filled or vacant d orbitals of the metal and gets adsorbed on the metal surfaces to form a protective layer. The aromatic ring carbon atom have a slight negative charge representing the aromatic character except the C atoms directly attached to hetero atoms that are electron deficient (i.e., they have positive charge) and therefore could be subject to nucleophilic attack. The charge distribution on the substituents atoms and also the number of active adsorption centres in each compound is similar to those followed in BEX and CIL.

References

[1] D. K. Verma, “Density Functional Theory (DFT) as a Powerful Tool for Designing Corrosion Inhibitors in Aqueous Phase,” in *Advanced Engineering Testing*, In Tech, 2018.

[2] P. Dohare, D. S. Chauhan, A. A. Sorour, and M. A. Quraishi, “DFT and experimental studies on the inhibition potentials of expired Tramadol drug on mild steel corrosion in hydrochloric acid,” *Mater. Discov.*, vol. 9, pp. 30–41, 2017.

[3] A. G. Bedir, M. Abd El-Raouf, S. Abdel-Mawgoud, N. A. Negm, and N. M. El Basiony, “Corrosion Inhibition of Carbon Steel in Hydrochloric Acid Solution Using Ethoxylated Nonionic Surfactants Based on Schiff Base: Electrochemical and Computational Investigations,” *ACS Omega*, vol. 6, no. 6, pp. 4300–4312, 2021.

[4] F. E. Abeng, V. C. Anadebe, P. Y. Nkom, K. J. Uwakwe, and E. G. Kamalu, “Experimental and theoretical study on the corrosion inhibitor potential of quinazoline derivative for mild steel in hydrochloric acid solution,” *J. Electrochem. Sci. Eng.*, vol. 11, no. 1, pp. 11–26, 2021.

[5] R. H. Albrakaty, N. A. Wazzan, and I. B. Obot, “Theoretical study of the mechanism of corrosion inhibition of carbon steel in acidic solution by 2-aminobenzothaizole and 2-Mercatobenzothiazole,” *Int. J. Electrochem. Sci.*, vol. 13, no. 4, pp. 3535–3554, 2018.

[6] H. I. Al-Shafey, R. S. Abdel Hameed, F. A. Ali, A. E. A. S. Aboul-Magd, and M. Salah, “Effect of expired drugs as corrosion inhibitors for carbon steel in 1M HCL solution,” *Int. J. Pharm. Sci. Rev. Res.*, vol. 27, no. 1, pp. 146–152, 2014.

[7] R. S. Abdel Hameed, E. A. Ismail, H. I. Al-Shafey, and M. A. Abbas, “Expired Indomethacin Therapeutics as Corrosion Inhibitors for Carbon Steel in 1.0 M Hydrochloric Acid Media,” *J. Bio- Tribo-Corrosion*, vol. 6, no. 4, 2020.

[8] G. K. Shamnamol, K. P. Sreelakshmi, G. Ajith, and J. M. Jacob, “Effective utilization of drugs as green corrosion inhibitor-A review,” in *AIP Conference Proceedings*, 2020, vol. 2225.

[9] M. A. Quraishi and D. S. Chauhan, "Drugs as Environmentally Sustainable Corrosion Inhibitors," *Sustain. Corros. Inhib. II Synth. Des. Pract. Appl. Part 1 - Drugs as Environ. Sustain. Corros. Inhib.*, pp. 1–17, 2021.

[10] A. S. Fouda, H. Ibrahim, S. Rashwaan, A. El-Hossiany, and R. M. Ahmed, "Expired drug (pantoprazole sodium) as a corrosion inhibitor for high carbon steel in hydrochloric acid solution," *Int. J. Electrochem. Sci.*, vol. 13, no. 7, pp. 6327–6346, 2018.

[11] A. Salhi *et al.*, "Keto-enol heterocycles as new compounds of corrosion inhibitors for carbon steel in 1 M HCl: Weight loss, electrochemical and quantum chemical investigation," *J. Mol. Liq.*, vol. 248, no. pp. 340–349, 2017.

[12] R. S. Abdel Hameed, A. H. Al-Bagawi, H. A. Shehata, A. H. Shamroukh, and M. Abdallah, "Corrosion Inhibition and Adsorption Properties of Some Heterocyclic Derivatives on C-Steel Surface in HCl," *J. Bio- Tribo-Corrosion*, vol. 6, no. 2, 2020.

[13] A. Srinivasulu and P. K. Kasthuri, "Study of inhibition and adsorption properties of mild steel corrosion by expired pharmaceutical Gentamicin drug in hydrochloric acid media," *Orient. J. Chem.*, vol. 33, no. 5, pp. 2616–2624, 2017.

[14] K. Ramya, K. K. Anupama, K. M. Shainy, and A. Joseph, "Corrosion protection of mild steel in hydrochloric acid solution through the synergistic of alkylbenzimidazoles and semicarbazide pair – Electroanalytical and computational studies," *Egypt. J. Pet.*, vol. 26, no. 2, pp. 421–437, 2017.

[15] M. A. M. El-Haddad, A. Bahgat Radwan, M. H. Sliem, W. M. I. Hassan, and A. M. Abdullah, "Highly efficient eco-friendly corrosion inhibitor for mild steel in 5 M HCl at elevated temperatures: experimental & molecular dynamics study," *Sci. Rep.*, vol. 9, no. 1, pp. 1–15, 2019.

[16] M. Ben Harb, S. Abubshait, N. Etteyeb, M. Kamoun, and A. Dhouib, "Olive leaf extract as a green corrosion inhibitor of reinforced concrete contaminated with seawater," *Arab. J. Chem.*, vol. 13, no. 3, pp. 4846–4856, 2020.

[17] S. K. Saha, M. Murmu, N. C. Murmu, I. B. Obot, and P. Banerjee, "Molecular level insights for the corrosion inhibition effectiveness of three amine derivatives on the carbon steel surface in the adverse medium: A combined density functional theory and molecular dynamics simulation study," *Surfaces and Interfaces*, vol. 10, no. March 2017, pp. 65–73, 2018.

[18] F. El-Hajjaji *et al.*, "Pyridazinium-based ionic liquids as novel and green corrosion inhibitors of carbon steel in acid medium: Electrochemical and molecular dynamics simulation studies," *J. Mol. Liq.*, vol. 249, pp. 997–1008, 2018.

[19] S. Ramkumar, D. Nalini, M. A. Quraishi, E. E. Ebenso, and C. Verma, "Anti-corrosive property of bioinspired environmental benign imidazole and isoxazoline heterocyclics: A cumulative studies of experimental and DFT methods," *J. Heterocycl. Chem.*, vol. 57, no. 1, pp. 103–119, 2020.

[20] E. M. Essassi, G. Aziate, H. Elmsellem, and I. Abdel-rahman, "Corrosion Inhibitor of Mild Steel Corrosion in Different Acidic Corrosion Inhibitor of Mild Steel Corrosion in Different Acidic Media (Hcl and H 2 So 4): Experimental and Quantum Chemical Study," vol. 318792491, no. 2017.

[21] A. K. Singh *et al.*, "Evaluation of anti-corrosion performance of an expired semi synthetic antibiotic cefdinir for mild steel in 1 M HCl medium: An experimental and theoretical study," *Results Phys.*, vol. 14, 2019.

[22] M. Ouakki *et al.*, "Quantum chemical and experimental evaluation of the inhibitory action of two imidazole derivatives on mild steel corrosion in sulphuric acid medium," *Heliyon*, vol. 5, no. 11, 2019.

[23] S. Kumar, D. Sharma, P. Yadav, and M. Yadav, "Experimental and quantum chemical studies on corrosion inhibition effect of synthesized organic compounds on N80 steel in hydrochloric acid," in *Industrial and Engineering Chemistry Research*, 2013, vol. 52, no. 39, pp. 14019–14029.

[24] S. Al-Baghdadi, T. S. Gaaz, A. Al-Adili, A. A. Al-Amiery, and M. S. Takriff, "Experimental studies on corrosion inhibition performance of acetylthiophene thiosemicarbazone for mild steel in HCl complemented with DFT investigation," *Int. J. Low-Carbon Technol.*, vol. 16, no. 1, pp. 181–188, 2021.

[25] P. Udhayakala, "Density functional theory calculations on corrosion inhibitory action of five azlactones on mild steel," *Available online www.jocpr.com J. Chem. Pharm. Res.*, vol. 6, no. 7, pp. 117–127, 2014.

[26] M. Marinescu, "Recent advances in the use of benzimidazoles as corrosion inhibitors," *BMC Chem.*, vol. 13, no. 1, pp. 1–21, 2019.

[27] E. Kowsari *et al.*, "In situ synthesis, electrochemical and quantum chemical analysis of an amino acid-derived ionic liquid inhibitor for corrosion protection of mild steel in 1M HCl solution," *Corros. Sci.*, vol. 112, pp. 73–85, 2016.

[28] M. Mobin, S. Zehra, and R. Aslam, "L-Phenylalanine methyl ester hydrochloride as a green corrosion inhibitor for mild steel in hydrochloric acid solution and the effect of surfactant additive," *RSC Adv.*, vol. 6, no. 7, pp. 5890–5902, 2016.

[29] A. Miralrio and A. E. Vázquez, "Plant extracts as green corrosion inhibitors for different metal surfaces and corrosive media: A review," *Processes*, vol. 8, no. 8, 2020.

[30] R. Mohan and A. Joseph, "Corrosion protection of mild steel in hydrochloric acid up to 313 K using propyl benzimidazole: Electroanalytical, adsorption and quantum chemical studies," *Egypt. J. Pet.*, vol. 27, no. 1, pp. 11–20, 2018.

[31] A. Mishra, C. Verma, H. Lgaz, V. Srivastava, M. A. Quraishi, and E. E. Ebenso, "Synthesis, characterization and corrosion inhibition studies of N-phenyl-benzamides on the acidic corrosion of mild steel: Experimental and computational studies," *J. Mol. Liq.*, vol. 251, pp. 317–332, 2018.

[32] P. Udhayakala and T. V Rajendiran, "A theoretical evaluation on benzothiazole derivatives as corrosion inhibitors on mild Steel," 2015.

[33] E. A. M. Gad, E. M. S. Azzam, and S. A. Halim, "Theoretical approach for the performance of 4-mercapto-1-alkylpyridin-1-ium bromide as corrosion inhibitors using DFT," *Egypt. J. Pet.*, vol. 27, no. 4, pp. 695–699, 2018.

[34] T. Arslan, F. Kandemirli, E. E. Ebenso, I. Love, and H. Alemu, "Quantum chemical studies on the corrosion inhibition of some sulphonamides on mild steel in acidic medium," *Corros. Sci.*, vol. 51, no. 1, pp. 35–47, 2009.

[35] K. R. Ansari *et al.*, "Macrocyclic compounds as green corrosion inhibitors for aluminium: Electrochemical, surface and quantum chemical studies," *Int. J. Corros. Scale Inhib.*, vol. 7, no. 3, pp. 443–459, 2018.

[36] I. Danaee, S. RameshKumar, M. RashvandAvei, and M. Vijayan, "Electrochemical and quantum chemical studies on corrosion inhibition performance of 2,2'-(2-hydroxyethylimino)bis[n-(alphaalpha-dimethylphenethyl)-n-methylacetamide] on mild steel corrosion in 1M HCL solution," *Mater. Res.*, vol. 23, no. 2, 2020.

Chapter VII

CHAPTER VII

ELECTROCHEMICAL INVESTIGATION ON THE PROTECTIVE DRUG/EPOXY COMPOSITES IN 3.5 % NaCl

7.1. Paint composite studies

7.1.1. Immersion test

Various paint composite of six coated mild steel plates (A-F) are immersed in 100ml of 3.5% of NaCl solution for 48 hours and the change at the surface was noted. The image of coated plate which was immersed in NaCl was shown below:

Figure 7.1. Corrosion inhibition in paint coated mild steel in 3.5% of NaCl before 48h

Figure 7.2. Corrosion inhibition in paint coated mild steel in 3.5% of NaCl before 48h

A-Bare mild steel

B-Paint coated mild steel (blank)

C-paint composite with Bexol coated mild steel

D--paint composite with Cilny coated mild steel

E--paint composite with Monit coated mild steel

F--paint composite with Prazopill coated mild steel

The observations of immersion test in 3.5% of NaCl was shown in table 7.1.From the observation made it is found that,

Sample A (bare mild steel) starts corroding in 3.5% of NaCl solution well before 12h of immersion and the metal dissolution increases with prolonged exposure after 48h [1]–[3]

Sample B (paint coated mild steel) is stable up to 36h but after which small blusters were formed on the surface. With further exposure, the paint starts to dissolve in the alkaline [4 - 5].

Sample C (paint composite with BEX coated mild steel) shows similar behaviour as that of Sample B

Sample D & E (paint composite with CIL &MON coated mild steel) does not undergo any Considerable change on exposure with 3.5% of NaCl up to 36h and after which only small blusters where formed on the surface at 48h of exposure [6]–[8].

Sample F (paint composite with PYR coated mild steel) is stable up to 24h and after that a small blusters are formed.

Thus it is confirmed that paint D and E gave good efficiency than others.

Table 7.1. The observations of immersion test in 3.5% of NaCl

Sample	After 12h	After 24h	After 36h	After 48h
A	Iron oxide formed	Iron oxide diffused into alkaline	Iron oxide diffused into alkaline	Iron oxide diffused into alkaline
B	No change	No change	Blusters are formed	Iron oxide diffused into alkaline
C	No change	No change	Blusters are formed on the surface	Small blusters formed on the surface
D	No change	No change	No change	Small blusters formed on the surface
E	No change	Small blusters formed on the surface	Small blusters formed on the surface	Iron oxide diffused into alkaline
F	No change	Small blusters formed on the surface	Small blusters formed on the surface	Iron oxide diffused into alkaline

7.1.2. Electrochemical studies

The electrochemical tests were carried out in a 3.5% of NaCl [9]–[11]. The surface area of coated plates exposed to alkaline solution [12]. It will serve as working electrode [13 - 14]

a) Bare mild steel

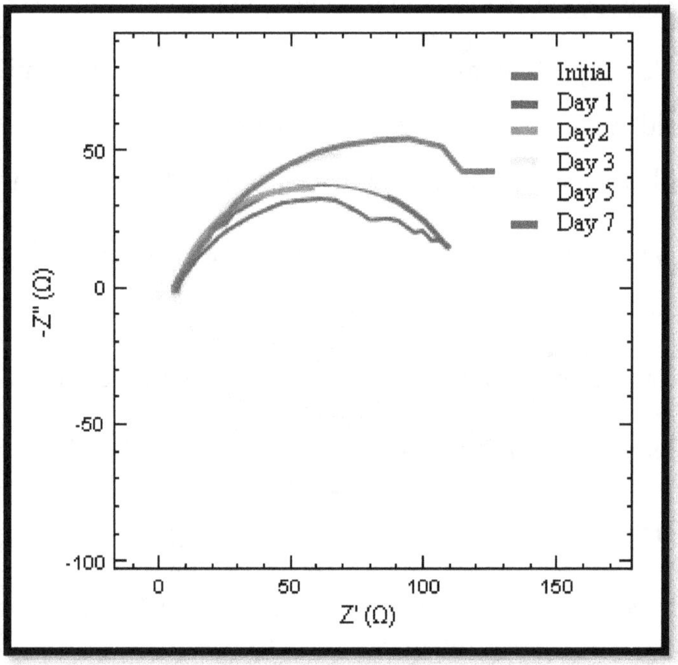

Figure 7.3. Nyquist plots for bare mild steel sample at different immersion times in 3.5% of NaCl

Table 7.2. Electrochemical Impedance parameters for corrosion bare mild steel in 3.5% of NaCl

Name of the inhibitor	Ageing period (Days)	R_s (Ω cm²)	C_{dl} (µF/cm²)	R_{ct} (Ω cm²)	R_c (Ω cm²)	R_{tot} (Ω cm²)
BARE	Initial	8.1	9 X 10^{-7}	3922	3849	7771
	Day 1	6.9	9 X 10^{-7}	2851	2748	5599
	Day 2	7.3	9 X 10^{-7}	854	774	1628
	Day 3	8.16	9 X 10^{-7}	512	412	924
	Day 5	4.8	9 X 10^{-7}	118	102	220
	Day 7	9.6	0.0000113	94	74	168

Nyquist plots of bare mild steel in 3.5% of NaCl at different time duration as shown in **Figure 7.5**.The kinetic parameters as Rs, Rct, Cdl are derived from these nyquist plots are specified in the **table 7.2.** [15 - 16]

Low interfacial double layer capacitance (Cdl), High charge transfer resistance (Rct) are related to corrosion resistance [17]. From the **table 7.2,** all the periods' of coatings Rct values are decreased. This decrease in Rct values shows that exposed to mild steel surface through cracking by galvanic action of paint coating [18]–[21]. Cdl values are increased which indicate that corrosion reaction occur at the surface.

R_{ct} values of bare mild steel for initial is 3922 Ω cm^2 and hence it will decreases the deposition time for first day it will decrease the Rct values from 3922 Ω cm^2 to 2851 Ω cm^2 and increase the immersion time [22]–[25].

b) Paint coated mild steel

The **paint** coated mild steel shows high resistance value for first day than other days. Its Nyquist plot was shown in **Figure 7.4** and the R_t values obtained at various time intervels from Day 1 to 7, is discussed in the **table 7.3** as follows.

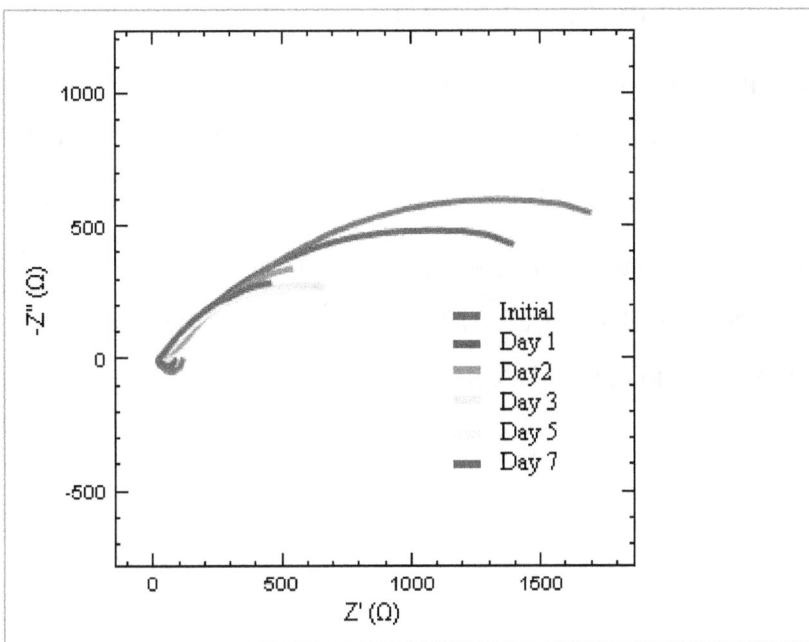

Figure 7.4.Nyquist plots for coated mild steel sample at different immersion times in 3.5% of NaCl

Initially Bexol coated sample shows higher total resistance value of 9585 Ωcm^2 which decreases to 265 Ωcm^2 (Day 7).This decrease may be due to the intrusion [26] of corrosion electrolyte on to the mild steel surface.

Table 7.3. Electrochemical Impedance parameters for corrosion of paint coated mild steel in 3.5% of NaCl

Name of the inhibitor	Ageing period (Days)	R_s (Ω cm^2)	C_{dl} ($\mu F/cm^2$)	R_{ct} (Ω cm^2)	R_c (Ω cm^2)	R_{tot} (Ω cm^2)
Paint coated mild steel	Initial	22.2	132	5034	4551	9585
	Day 1	37.5	122	2155	2026	4181
	Day 2	26.4	$9 X 10^{-7}$	834	535	1369
	Day 3	22.2	132	239	151	390
	Day 5	45	$9 X 10^{-7}$	176	168	344
	Day 7	33.8	$9 X 10^{-7}$	85	180	265

c) Paint composite with BEX coated mild steel

The **paint** composite with BEX coated mild steel shows high resistance value for first day than other days. Its Nyquist plot was shown in **Figure 7.7** and the R_t values obtained at various time intervels from Day 1 to 7, is discussed in the table **7.4** as follows.

Initially Bexol coated sample shows higher total resistance value of 9585 Ωcm^2 which decreases to 265 Ωcm^2 (Day 7).This decrease may be due to the intrusion of corrosion electrolyte on to the mild steel surface.

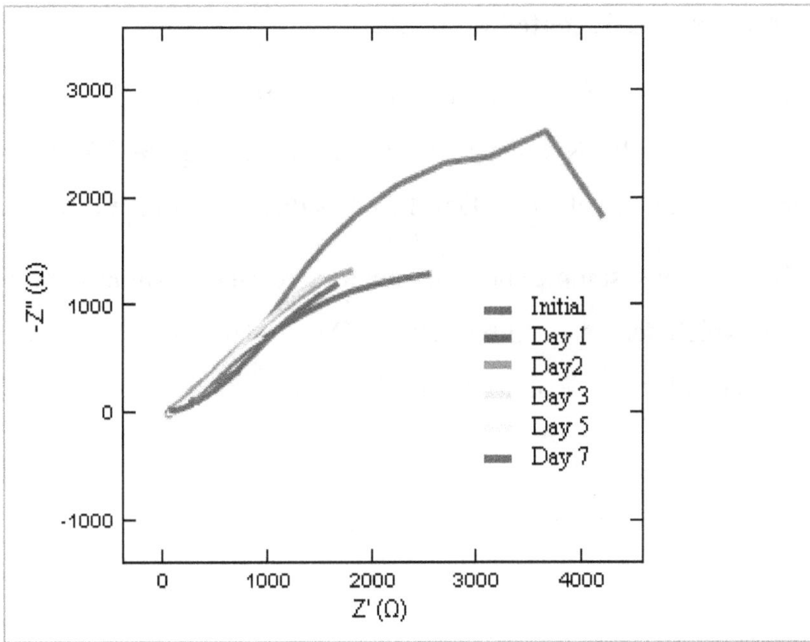

Figure 7.5. Nyquist plots for paint with BEX coated mild steel sample at different immersion times in 3.5% of NaCl.

Table 7.4. Electrochemical Impedance parameters for corrosion of BEX with paint coated mild steel in 3.5% of NaCl

Name of the inhibitor	Ageing period (Days)	R_s(Ω cm^2)	C_{dl} (μF/cm^2)	R_{ct} (Ω cm^2)	R_c(Ω cm^2)	R_{tot} (Ω cm^2)
Bexol	Initial	58	118	3830	3285	7115
	Day 1	113	104	6947	3781	10728
	Day 2	42.6	213	1600	1092	2692
	Day 3	68	316	1529	989	2518
	Day 5	57	533	885	633	1518
	Day 7	93	.0203	472	335	807

d) Paint composite with CIL coated mild steel

The **paint** composite with CIL coated mild steel shows high resistance value for first day than other days. Its Nyquist plot was shown in **Figure 7.6** and the R_t values obtained at various time intervels from Day 1 to 7, is discussed in the **table 7.5** as follows.

Initially CIL coated sample shows higher total resistance value at initially increases from 8365 Ωcm^2 which decreases to 623 Ωcm^2 (Day 7).This decrease may be due to the intrusion of corrosion electrolyte on to the mild steel surface.

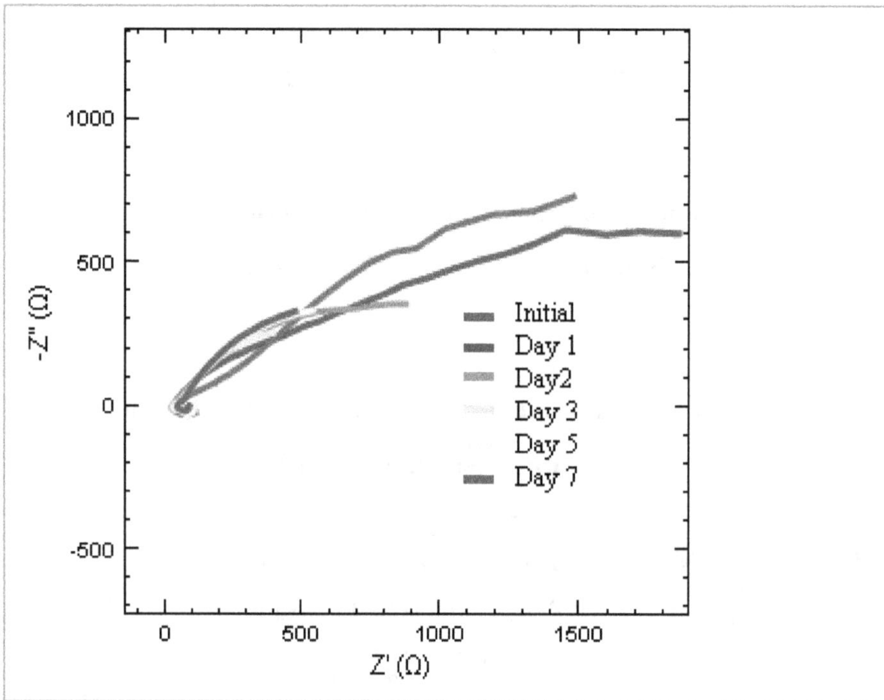

Figure 7.6. Nyquist plots for paint with CIL coated mild steel sample at different immersion times in 3.5% of NaCl.

Table 7.5. Electrochemical Impedance parameters for corrosion of CILNY with paint coated mild steel in 3.5% of NaCl

Name of the inhibitor	Ageing period (Days)	R_s (Ω cm^2)	C_{dl} (μF/cm^2)	R_{ct}(Ω cm^2)	R_c(Ω cm^2)	R_{tot} (Ω cm^2)
Cilny	Initial	39	92.8	4045	4320	8365
	Day 1	61	22.5	3052	3181	6233
	Day 2	32	9 X 10^{-7}	1229	1048	2277
	Day 3	31	9 X 10^{-7}	903	324	1227
	Day 5	48	1.15	532	126	658
	Day 7	48.7	1.28	527	96	623

e) Paint composite with MON coated mild steel

The **paint** composite with MON coated mild steel shows high resistance value for first day than other days. Its Nyquist plot was shown in **Figure 7.6** and the R_t values obtained at various time intervels from Day 1 to 7, is discussed in the **table 7.6** as follows.

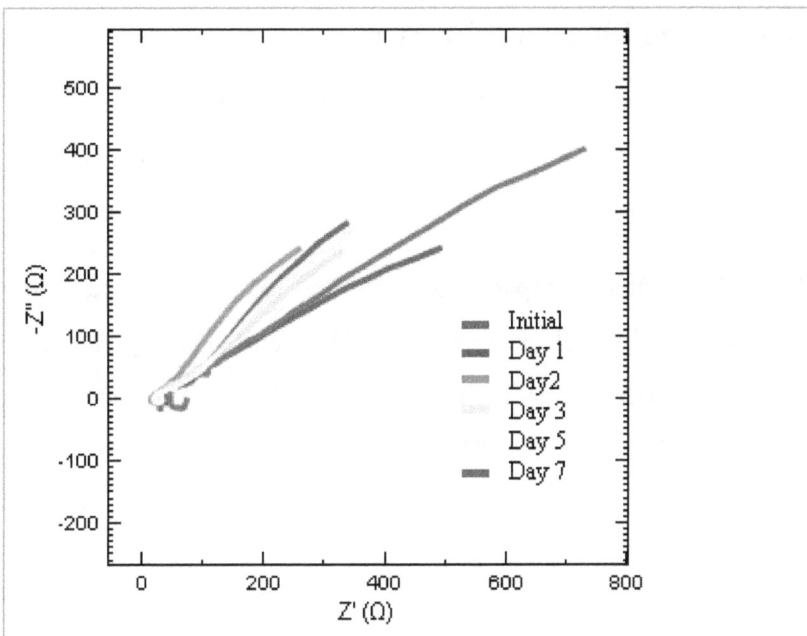

Figure 7.7. Nyquist plots for paint with MON coated mild steel sample at different immersion times in 3.5% of NaCl.

Initially MON coated sample shows higher total resistance value at initially increases from 9862 Ωcm^2 which decreases to 4717 Ωcm^2 (Day 7).This decrease may be due to the intrusion of corrosion electrolyte on to the mild steel surface.

Table 7.6. Electrochemical Impedance parameters for corrosion of MONIT with paint coated mild steel in 3.5% of NaCl

Name of the inhibitor	Ageing period (Days)	R_s (Ω cm^2)	C_{dl} (μF/cm^2)	R_{ct} (Ω cm^2)	R_c (Ω cm^2)	R_{tot} (Ω cm^2)
Monit	Initial	1530	0.595	6552	3310	9862
	Day 1	122	170	2411	2306	4717
	Day 2	48	571	1017	1098	2115
	Day 3	46	9 X 10^{-7}	825	711	1536
	Day 5	64	0.00574	578	271	849
	Day 7	59	9 X 10^{-7}	265	112	377

f) Paint composite with PYR coated mild steel

The **paint** composite with PYR coated mild steel shows high resistance value for first day than other days. Its Nyquist plot was shown in **Figure 7.6** and the R_t values obtained at various time intervels from Day 1 to 7, is discussed as follows.

Initially PYR coated sample shows higher total resistance value at initially increases from 7193 Ωcm^2 which decreases to 6933 Ωcm^2 (Day 7).This decrease may be due to the intrusion of corrosion electrolyte on to the mild steel surface.

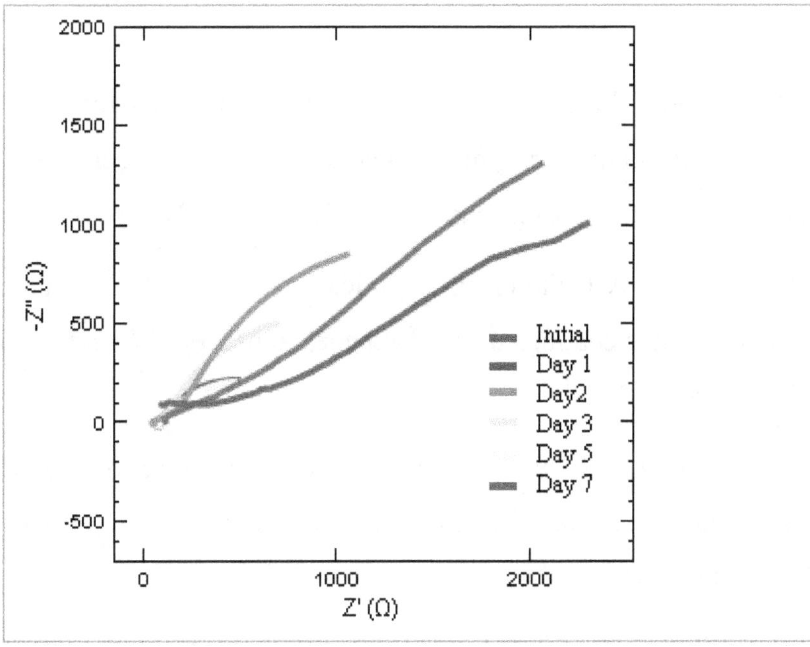

Figure 7.8. Nyquist plots for paint with PYR coated mild steel sample at different immersion times in 3.5% of NaCl.

Table 7.7. Electrochemical Impedance parameters for corrosion of PYR with paint coated mild steel in 3.5% of NaCl

Name of the inhibitor	Ageing period (Days)	R_s (Ω cm^2)	C_{dl} (μF/cm^2)	R_{ct} (Ω cm^2)	R_c (Ω cm^2)	R_{tot} (Ω cm^2)
Prazopill	Initial	36.4	61.6	3495	3698	7193
	Day 1	51	9×10^{-7}	4809	2124	6933
	Day 2	17	9×10^{-7}	492	387	879
	Day 3	21	0.000922	337	213	550
	Day 5	24	9×10^{-7}	127	133	260
	Day 7	23	0.00006	98	74	172

References

[1] A. U. Ammar, M. Shahid, M. K. Ahmed, M. Khan, A. Khalid, and Z. A. Khan, "Electrochemical study of polymer and ceramic-based nanocomposite coatings for corrosion protection of cast iron pipeline," *Materials (Basel).*, vol. 9, no. 3, 2018.

[2] S. C. Sahu *et al.*, "A facile electrochemical approach for development of highly corrosion protective coatings using graphene nanosheets," *Electrochem. commun.*, vol. 32, pp. 22–26, 2013.

[3] M. K. Ahmed *et al.*, "Electrochemical comparison of SAN/PANI/FLG and ZnO/GO coated cast iron subject to corrosive environments," *Materials (Basel).*, vol. 11, no. 11, 2018.

[4] N. Hammouda, H. Chadli, G. Guillemot, and K. Belmokre, "The Corrosion Protection Behaviour of Zinc Rich Epoxy Paint in 3% NaCl Solution," *Adv. Chem. Eng. Sci.*, vol. 01, no. 02, pp. 51–60, 2011.

[5] N. Hammouda, G. Guillemot, and K. Belmokre, "The Corrosion Protection Behaviour of Zinc Rich Epoxy Paint in 3 % NaCl Solution," vol. c, pp. 51–60, 2011.

[6] Z. Zhang *et al.*, "Molecule adsorption and corrosion mechanism of steel under protection of inhibitor in a simulated concrete solution with 3.5% NaCl," *RSC Adv.*, vol. 8, no. 37, pp. 20648–20654, 2018.

[7] A. Pradityana, Sulistijono, A. Shahab, L. Noerochim, and D. Susanti, "Inhibition of Corrosion of Carbon Steel in 3.5% NaCl Solution by Myrmecodia Pendans Extract," *Int. J. Corros.*, vol. 2016, 2016.

[8] R. Hsissou, B. Benzidia, N. Hajjaji, and A. Elharfi, "Elaboration and electrochemical studies of the coating behavior of a new nanofunctional epoxy polymer on E24 steel in 3.5 % NaCl," *Port. Electrochim. Acta*, vol. 36, no. 4, pp. 259–270, 2018.

[9] Y. Liu, H. Li, and Z. Li, "EIS Investigation and Structural Characterization of Different Hot-Dipped Zinc-Based Coatings in 3 . 5 % NaCl Solution," vol. 8, pp. 7753–7767, 2013.

[10] F. A. Ghauri, M. A. Raza, M. S. Baig, and S. Ibrahim, "Corrosion study of the graphene oxide and reduced graphene oxide-based epoxy coatings," *Mater. Res. Express*, vol. 4, no. 12, pp. 0–21, 2017.

[11] Z. Mirzakhanzadeh, A. Kosari, M. H. Moayed, R. Naderi, P. Taheri, and J. M. C. Mol, "Enhanced corrosion protection of mild steel by the synergetic effect of zinc aluminum polyphosphate and 2-mercaptobenzimidazole inhibitors incorporated in epoxy-polyamide coatings," *Corros. Sci.*, vol. 138, no. June 2017, pp. 372–379, 2018.

[12] B. Ramezanzadeh, S. Niroumandrad, A. Ahmadi, M. Mahdavian, and M. H. Mohamadzadeh Moghadam, "Enhancement of barrier and corrosion protection performance of an epoxy coating through wet transfer of amino functionalized graphene oxide," *Corros. Sci.*, vol. 103, pp. 283–304, 2016.

[13] N. Parhizkar, T. Shahrabi, and B. Ramezanzadeh, "A new approach for enhancement of the corrosion protection properties and interfacial adhesion bonds between the epoxy coating and steel substrate through surface treatment by covalently modified amino functionalized graphene oxide film," *Corros. Sci.*, vol. 123, pp. 55–75, 2017.

[14] K. Qi, Y. Sun, H. Duan, and X. Guo, "A corrosion-protective coating based on a solution-processable polymer-grafted graphene oxide nanocomposite," *Corros. Sci.*, vol. 98, pp. 500–506, 2015.

[15] N. Raghavendra, "Expired abacavir sulfate drug as non-toxic corrosion inhibitor for mild steel (MS) in 3 M hydrochloric acid system," *Gazi Univ. J. Sci.*, vol. 32, no. 4, pp. 1113–1121, 2019.

[16] N. Raghavendra, "Expired abacavir sulfate drug as non-toxic corrosion inhibitor for mild steel (MS) in 3 M hydrochloric acid system," *Gazi Univ. J. Sci.*, vol. 32, no. 4, pp. 1113–1121, 2019.

[17] A. S. Fouda, H. S. El-Desoky, M. A. Abdel-Galeil, and D. Mansour, "Niclosamide and dichlorphenamide: new and effective corrosion inhibitors for carbon steel in 1M HCl solution," *SN Appl. Sci.*, vol. 3, no. 3, 2021.

[18] V. Rajasekharan, T. Stalin, S. Viswanathan, and P. Manisankar, "Electrochemical evaluation of anticorrosive performance of organic acid doped polyaniline based coatings," *Int. J. Electrochem. Sci.*, vol. 8, no. 9, pp. 11327–11336, 2013.

[19] Z. Yu *et al.*, "Preparation of graphene oxide modified by titanium dioxide to enhance the anti-corrosion performance of epoxy coatings," *Surf. Coatings Technol.*, vol. 276, pp. 471–478, 2015.

[20] O. Dagdag, M. Galai, M. Ebn Touhami, A. Essamri, and A. El Harfi, "Electrochemical study of the polymer behavior of an epoxy coating on carbon steel in 3% NaCl using polarization curves and SIE," *J. Mater. Environ. Sci.*, vol. 7, no. 9, pp. 3454–3464, 2016.

[21] K. Karattu Veedu, T. Peringattu Kalarikkal, N. Jayakumar, and N. K. Gopalan, "Anticorrosive Performance of Mangifera indica L. Leaf Extract-Based Hybrid Coating on Steel," *ACS Omega*, vol. 4, no. 6, pp. 10176–10184, 2019.

[22] "Advanced Electrochemical Methods for Characterizing The Performance of Organic Coatings North Dakota State University," 2012.

[23] G. C. E. Arguelles, M. C. D. Torres, B. J. U. Wu, G. Jerry, and R. C. Ong, "(Calamansi) Citrofortunella Microcarpa Rind Extract As Plant Derived Green Corrosion Inhibitor For Mild Steel In 1. 0M Hcl Microcarpa Rind Extract As Plant Derived Green Corrosion Inhibitor For Mild Steel In 1 . 0M HCl," 2020.

[24] B. Nikpour, B. Ramezanzadeh, G. Bahlakeh, and M. Mahdavian, "Synthesis of graphene oxide nanosheets functionalized by green corrosion inhibitive compounds to fabricate a protective system," *Corros. Sci.*, vol. 127, pp. 240–259, 2017.

[25] Z. Chen, C. He, F. Yu, and Y. Wang, "Study and application of electrochemical impedance spectroscopy for quickly evaluating the performance of coatings and predicting the failure time in the development of waterborne epoxy micaceous iron oxide coatings," *Int. J. Electrochem. Sci.*, vol. 12, no. 4, pp. 2798–2812, 2017.

[26] E. B. Caldona *et al.*, "Corrosion inhibition of mild steel in acidic medium by simple azole-based aromatic compounds," *J. Electroanal. Chem.*, vol. 880, p. 114858, 2021.

Chapter VIII

CHAPTER VIII

SUMMARY AND CONCLUSION

In the present study, Electrochemical and non-electrochemical properties of drug four inhibitors namely Bexol, Cilny, Monit, and Prazopill in relative to inhibition mechanism on mild steel in 1M HCl and 3.5 % NaCl media have been studied. The studies that have been carried out were I. Weight loss studies II. Electrochemical studies III. Quantum chemical Studies IV. Paint composite Electrochemical studies (EIS)

8.1. Weight loss measurements

1. The inhibition efficiency of the inhibitors (BEX, CIL, MON & PRY) with different concentrations **(5 – 60 mg/L)** on mild steel in 1M HCl has been evaluated after different hours of immersion (3- 24 hours).
2. The inhibition efficiency for all the investigated pharmaceutically active heterocyclic drugs increased with an increase in concentration from 3h to 24h and then decreased for 24 h. But, with an increase in temperature from 30°C - 70°C, I.E (%) decreases.
3. Each inhibitor shows a maximum efficiency at the optimum concentration of 50 mg/L.
4. On further increase in concentration to about 60mg/L, there is no further increase in the efficiency of the studied drugs.

8.2. Electrochemical studies

1. The decrease in C_{dl} and increase in R_{ct} for all the studied inhibitors may be due to the gradual replacement of water molecules by the adsorbed inhibitor
2. I_{corr} decreased with an increase in the concentration of inhibitor in the order HCl which reiterates that the inhibitor is more effective in HCl.

8.3. Quantum chemical studies

1. **BEX and PYR** have one piperadyl and piperazine ring, **MON** contains 2 NO_2 groups as substituents at the propane ring. **CIL** contains NH substituent at the benzene ring and one dicarboxylate group attached to the main ring.
2. The C-C bond length of benzene rings ranges between 1.39-1.42 for all the studied drugs.
3. BEX has the lowest energy gap among all the drugs. Hence it is expected to show a greater inhibitive action than other investigated drugs.
4. **BEX, CIL, and PYR** show lower values of electronegativity than **MON** possessing a higher value of electronegativity.
5. Mulliken charge distribution and Fukui function show that **N and O atoms** carry the highest negative charges. These atoms are therefore likely to readily donate electrons to the partially filled or vacant d orbitals of the metal and get adsorbed on the metal surfaces to form a protective layer.

8.4. Paint composite electrochemical studies

1. **Bare mild steel:** R_{ct} values of bare mild steel decrease with an increase in immersion time in 3.5 % NaCl solution.
2. **Paint coated mild steel containing inhibitors:** Mild steel substrates coated with inhibitors showed a slow decrease in R_{ct} when compared to that of bare steel substrate which may be due to the slow ingression of corrosion electrolyte into the coating surface. This shows that the inhibitor molecules present in the coating mixture provide a protective barrier.

By consolidating the above results, it can be concluded that all four drugs inhibited the corrosion process on mild steel through the physisorption mechanism. The inhibitor adsorption on the mild steel surface is spontaneous as ΔG_{ads} were negative and support Langmuir and Temkin adsorption models. All the inhibitors were found to be best in corrosion inhibition of mild steel in 1M HCl and 3.5 wt. % NaCl solution. From Fukui indices calculations, Mulliken charges on the atoms of the inhibitors, it is clear that Nitrogen and Oxygen atoms were the sites for an electrophilic attack on the studied drugs.

www.ingramcontent.com/pod-product-compliance
Lightning Source LLC
LaVergne TN
LVHW082333200926
848481LV00012B/3334
* 9 7 9 8 2 3 2 1 1 6 7 3 6 *